Cambridge Elements ≡

Elements in the Philosophy of Mathematics
edited by
Penelope Rush
University of Tasmania
Stewart Shapiro
The Ohio State University

SEMANTICS AND THE ONTOLOGY OF NUMBER

Eric Snyder
Ashoka University

CAMBRIDGE
UNIVERSITY PRESS

CAMBRIDGE
UNIVERSITY PRESS

University Printing House, Cambridge CB2 8BS, United Kingdom

One Liberty Plaza, 20th Floor, New York, NY 10006, USA

477 Williamstown Road, Port Melbourne, VIC 3207, Australia

314–321, 3rd Floor, Plot 3, Splendor Forum, Jasola District Centre, New Delhi – 110025, India

79 Anson Road, #06–04/06, Singapore 079906

Cambridge University Press is part of the University of Cambridge.

It furthers the University's mission by disseminating knowledge in the pursuit of education, learning, and research at the highest international levels of excellence.

www.cambridge.org
Information on this title: www.cambridge.org/9781108456258
DOI: 10.1017/9781108602259

First published 2021

A catalogue record for this publication is available from the British Library.

ISBN 978-1-108-45625-8 Paperback
ISSN 2399-2883 (online)
ISSN 2514-3808 (print)

Semantics and the Ontology of Number

Elements in the Philosophy of Mathematics

DOI: 10.1017/9781108602259
First published online: April 2021

Eric Snyder
Ashoka University
Author for correspondence: Eric Snyder, eric.snyder@ashoka.edu.in

Abstract: What are the meanings of number expressions, and what can they tell us about questions of central importance to the philosophy of mathematics, specifically 'Do numbers exist?' This Element attempts to shed light on this question by outlining a recent debate between substantivalists and adjectivalists regarding the semantic function of number words in numerical statements. After highlighting their motivations and challenges, I develop a comprehensive polymorphic semantics for number expressions. I argue that accounting for their numerous meanings and how they are related leads to a strengthened argument for realism, one which renders familiar forms of nominalism highly implausible.

Keywords: semantics of number expressions, realism and nominalism, substantivalism and adjectivalism, counting and measuring, the easy argument for numbers

ISBNs: 9781108456258 (PB), 9781108602259 (OC)
ISSNs: 2399-2883 (online), 2514-3808 (print)

Contents

1 Introduction

One of the central problems within contemporary philosophy of mathematics is BENACERRAF'S DILEMMA, raised by Paul Benacerraf (1973).[1] In effect, Benacerraf charges that we cannot have both an empirically respectable semantics for number words and an empirically respectable epistemology of arithmetic. More exactly, Benacerraf argues that because (1a,b) have the same underlying "logical form," or semantic representation, given in (1c), '17' in (1b) is a NUMERAL (a name of a number), similar to 'New York.'

(1)　a.　There are at least three large cities older than New York.

　　　b.　There are at least three perfect numbers greater than 17.

　　　c.　There are at least three Fs which bear R to a.

If so, then because (1b) is true, that numeral requires a referent. Moreover, the most obvious candidate referent is a *number*, which is presumably an abstract object. Hence, making semantic sense of basic arithmetic statements requires positing an ontology of abstract objects.

On the other hand, Benacerraf requires a CAUSAL THEORY of epistemology, whereby a knowledge report of the form 'S knows that p' is true only if there is a causal chain linking S and what that knowledge is about. In the case of (1a,b), for instance, a causal theory would require a causal chain linking us to New York and the number seventeen, respectively. However, the latter is inconsistent with numbers being abstract: since abstract objects exist outside of space and time, they are causally inefficacious. So, if knowledge of the kind expressed by (1b) requires a causal link, then it would appear that arithmetic knowledge is generally impossible.

The traditional response to Benacerraf's Dilemma has been to reject the epistemic horn: even if arithmetic statements like (1b) refer to abstract objects, knowing what is expressed by such statements does not require a causal link between agents and those objects.[2] However, an interesting alternative strategy has emerged in recent times. According to it, we should question the primary assumption driving Benacerraf's semantic horn, namely that '17' in (1b) is a referring term. If it is instead a *non-referential* expression, then explaining the truth of (1b) may not require postulating numbers.[3]

---

[1] Not to be confused with another famous dilemma posed by Benacerraf (1965), known as Benacerraf's Identification Problem.

[2] See e.g. Shapiro (1997).

[3] See e.g. Nutting (2018).

This raises an important question. Apart from the vague intuition that (1a,b) have the same "logical form," why think that number words *ever* function referentially within true mathematical statements? Bob Hale (1987) offers one influential argument, the first premise of which is:[4]

Successful Reference: If a range of expressions functions as singular terms in true statements, then there are objects denoted by expressions belonging to that range.

By definition, singular terms are expressions whose express semantic function is to refer, presumably to objects. Prototypical examples include names like 'Mars' in (2a) and definites like 'the planet Mars' in (2b).

(2)   a. Mars is a red planet.

     b. The planet Mars is red.

Now compare (2a,b) with (3a,b).

(3)   a. Two is an even number.

     b. The number two is even.

Given the surface syntactic similarities between (2a,b) and (3a,b), it would thus appear difficult to deny that the underlined expressions in (3a,b) are singular terms if those in (2a,b) are singular terms.

Hence the motivation for Hale's second premise:

Singular Terms: Numerals, and many other numerical expressions besides, function as singular terms in many true statements (of both pure and applied mathematics).

By statements of "pure mathematics," Hale presumably has in mind those characteristic of a particular branch of mathematics, such as number theory. These may include formal statements of the theory, or their informal natural language counterparts. For example, (4a,b) might count as "pure" statements of number theory.

(4)   a. $\forall x \in \mathbb{N}.\exists y \in \mathbb{N}.\, S(x) = y$

     b. Every natural number has a successor.

(3a,b) might also count as "pure statements," since they report something provable within number theory. In contrast, (5a,b) would presumably count as statements of "applied mathematics," as they report something involving a potential *application* of numbers, namely counting.

---

[4] Successful Reference, along with Singular Terms, Candidacy, and Existence below, are near direct quotations from Hale (1987).

(5) a. Mars has exactly two moons.

b. The number of Mars's moons is two.

Note that (5a,b) are not only plausibly equivalent, but also that 'two' seemingly functions as a singular term in (5b). Thus, not just statements of "pure arithmetic" are potentially relevant to Singular Terms.

Hale's final premise is:

Candidacy: Numbers are the best candidates to play the role of the referents of numerical expressions used referentially.

Intuitively, basic arithmetic is *about* numbers. Thus, insofar as examples like (3a,b) and (5b) are statements of pure and applied mathematics, broadly construed, they too are presumably *about* numbers. Furthermore, it is hard to see how (3a) could be true without Candidacy. Indeed, if 'two' referred to something other than a number, then (3a) would be false, presumably.

Jointly, Hale's premises imply:

Existence: Therefore, there exist objects denoted by numerical expressions, namely numbers.

Suppose (3a,b) are true. Then, by Successful Reference and Singular Terms, there is an object referenced by 'two' and 'the number two.' By Candidacy, that object is a number. So, by Existence, there exists an object which is the number two. This is just a restatement of REALISM, or the view that numbers exist. Accordingly, call this HALE'S ARGUMENT FOR REALISM.

To clarify terms, let's call numerical statements in which number expressions appear to function as singular terms referring to numbers, such as (1b), (3a,b), and (5b), APPARENTLY REFERENTIAL NUMERICAL STATEMENTS (ARNSs), and apparent singular terms featuring in those statements APPARENT SINGULAR TERMS (ASTs). [By NUMERICAL STATEMENTS, I mean utterances of sentences purporting to be about or otherwise feature number. And by NUMBER EXPRESSIONS, I mean simple or complex expressions purporting to be about or otherwise feature number, including 'two' in (3a) and (5a), 'the number two,' and 'the number of Mars's moons.'] Hale's Argument highlights two questions vital to the relationship between semantics and the ontology of number. First, are ARNSs true? Secondly, are ASTs featuring in ARNSs actually singular terms?

Three influential views within the philosophy of mathematics emerge depending on how these questions are answered. According to what I call REFERENTIALISM, ARNSs are true, and ASTs are singular terms. Versions of referentialism have been defended by e.g. Frege (1884), Shapiro (1997), and Hale and Wright (2001). According to what I call NON-REFERENTIALISM, ARNSs are true, but ASTs are not actually singular terms. Though various philosophers have proposed analyses suggesting non-referentialism, it has arguably been defended most forcefully and extensively by Hofweber (2005, 2007, 2016). Finally, according to what I call ERROR THEORY, although ASTs are singular terms, ARNSs are not actually true. Versions of error theory have been defended by e.g. Field (1980) and Leng (2005).

Clearly, if referentialism is correct, then so is Singular Terms. Thus, referentialism supports realism through Hale's Argument. In contrast, non-referentialism and error theory are both consistent with NOMINALISM—the view that numbers do not exist—though for importantly different reasons. This plausibly reflects a difference in philosophical methodology: whereas referentialists and non-referentialists agree that the question of whether numbers exist is to be determined largely, if not primarily, by the best available linguistic evidence, error theorists typically presuppose nominalism, for non-linguistic reasons. Thus, the dispute between referentialists and non-referentialists concerns the linguistic facts: Do they, or don't they, reveal ASTs to be genuinely referential? In contrast, error theorists typically take surface syntactic appearances at face value: ASTs really are referential. However, since their would-be referents do not exist, ARNSs are not true.

This short monograph is principally concerned with the referentialism/non-referentialism debate. The primary question to be addressed is whether, given the best available linguistic evidence, ASTs are singular terms referring to numbers. The connection to Benacerraf's Dilemma is immediate: if referentialism is correct, then the truth of ARNSs would seemingly straightforwardly vindicate realism, and with it Benacerraf's semantic horn. But if non-referentialism is correct, then not only is the truth of ARNSs consistent with nominalism, we would appear to have a novel and sufficiently general way out of Benacerraf's Dilemma.

Although my principal target here is the empirical motivation for referentialism and non-referentialism, this should also be of significant interest to error theorists. Even if the primary motivations for adopting nominalism are non-linguistic, whether ASTs are actually referential is an empirical matter. In this respect, one crucial aspect of error theory depends directly on the empirical viability of referentialism. Furthermore, if non-referentialism is correct, then there is no apparent need to claim that ARNSs are not true, contrary to intuition and basic mathematics. In this respect, non-referentialism would appear preferable to error theory, and so its other aspect plausibly depends indirectly on the empirical viability of non-referentialism.

Ultimately, I will argue here that the semantic evidence strongly supports referentialism, but for reasons which have been largely unappreciated within the philosophical literature. There are two notable tendencies among philosophers when discussing the relationship between numerical discourse and ontology. The first is to focus on a small handful of "mathematical" examples. Often, these include just statements of "pure mathematics," such as '2 + 2 = 4' or (4a,b). When other kinds of examples are discussed, these are often limited to statements of "applied mathematics" like (5a,b). The second tendency is to assume that number words function exclusively either as singular terms, or else as non-referential expressions. Thus, in cases like (5a,b), where 'two' appears to function *both* referentially and non-referentially, surface syntactic appearances must be misleading.

I believe that both tendencies are mistaken, and that recognizing this is crucial to understanding the ontological ramifications of numerical discourse. First, number words have *many* kinds of uses beyond those discussed above. For example, 'two' can be used not only as a numeral *and* as a quantificational expression, as witnessed in (5a,b), but also as a predicate and a modifier. Furthermore, it can be used to convey different kinds of numerical information—cardinal, arithmetic, ordinal, and measurement—each corresponding to different potential applications of numbers—counting, calculating, ordering, and measuring. Ultimately, an empirically adequate semantics for number words should not only explain how number words can be used in these various ways, but also, and perhaps more importantly, how the different meanings expressed by these various uses are *related*. And a semantics for number words can inform the question of whether numbers exist, presumably, only if it is empirically adequate.

My contention is that the only kind of semantic analysis capable of explaining these features overwhelmingly supports referentialism, and thus realism. More exactly, such an analysis will not only recognize that number words can take on a wide variety of different semantic functions and corresponding meanings, but also that these are related in virtue of sharing a certain element in common, namely a *number*. This in turn provides novel, empirical support for Singular Terms and Candidacy, suitably understood, thus resulting in a strengthened form of Hale's Argument. A philosophically significant, and perhaps surprising, consequence will be that *all* numerical discourse presupposes an ontology of numbers. This includes not just cases like (3a,b), which involve explicit reference to numbers, but also those like (5a), which do not. As a result, if numbers do not exist, then *none* of our numerical discourse is true—not just overtly "mathematical" statements, like (1b) or (3a,b), but even more mundane statements like (5a).

The rest of the monograph is divided into two sections. Section 2 outlines the empirical motivations for two influential strategies within the philosophical literature for dealing with examples like (5a,b), namely the substantival strategy and the adjectival strategy. Both promote or else are a form of referentialism or non-referentialism, and both can be deployed to address a particular instance of Hale's Argument, namely the Easy Argument for Numbers. I will argue that extant versions of both strategies face significant theoretical and empirical challenges.

In Section 3, I argue that both strategies are mistaken in virtue of wrongly assuming that number words featuring in examples like (5a,b) are exclusively referential or exclusively non-referential. After observing that number words have a wide variety of both referential and non-referential uses, I sketch a comprehensive semantics which not only provides meanings appropriate for all of these uses, but which also explains how those meanings are systematically related. I then explain how the resulting semantics affords a strengthened version of Hale's Argument, thereby undermining non-referentialism while also rendering error theory highly implausible.

Although this monograph is principally targeted at an ontological debate within the philosophy of mathematics, I hope that it may be potentially valuable to other areas of philosophy, and perhaps even linguistic semantics. There are analogous debates in other areas of philosophy, notably with respect to proper names and natural kind terms, and I think certain methodological points raised here carry over to these other areas. And while the semantics developed is by no means revolutionary, it does raise important empirical questions which should be of independent linguistic interest.

That said, there are numerous fascinating theoretical and empirical issues I have needed to ignore, for space. These include a more complete discussion of nominalist programs within the philosophy of mathematics, possible rejoinders to my arguments, empirical matters relevant to deciding the lexical meanings of number words, and, perhaps most obviously, how best to deal with Benacerraf's Dilemma in light of my arguments for referentialism. I intend to return to these matters in future work.

2 Substantivalism and Adjectivalism

As mentioned, this monograph is principally concerned with a recent debate regarding the meanings of number expressions and their ontological import. Specifically, it centers on Hale's Argument. The claim is that if the underlined expressions in (6a-c) are singular terms referring to numbers, then the truth of (6a-c) implies that numbers exist.

(6)   a.   <u>Four</u> is an even number.

   b.   <u>The number four</u> is even.

   c.   The number of Jupiter's moons is <u>four</u>.

But what exactly *is* a singular term, anyway, and why should we think that the underlined expressions in (6a-c) are among them? To this end, one might follow Dummett (1973) in characterizing singular terms via the kinds of inferences they (and only they) license. For example, whereas prototypical singular terms, such as names, apparently license the intuitive entailments in (7), prototypical non-referential expressions, such as 'nobody' or 'most Americans,' apparently do not.

(7)　a.　Mary is at least 18 years old or younger ⊨ Mary is at least 18 years old or Mary is younger than 18

　　　b.　Mary is intelligent ⊨ Someone is intelligent

　　　c.　Mary is female and Mary supports legalizing marijuana ⊨ Someone is female and supports legalizing marijuana

Note that numerals and definites like 'the number four' pattern similarly to names in this respect.

(8)　a.　{Four/The number four} is greater than or equal to five or less than five ⊨ {Four/The number four} is greater than or equal to five or {four/the number four} is less than five

　　　b.　{Four/The number four} is even ⊨ Some number is even

　　　c.　{Two/The number two} is even and {two/the number two} is prime ⊨ Some number is even and prime

Something similar can be said for (6c), it seems.

(9)　The number of Jupiter's moons is four ⊨ The number of Jupiter's moons is something, namely four

So, if licensing these entailments suffices for singular termhood,[5] examples like (6a-c) would seemingly vindicate Singular Terms.

This section is organized around the apparent entailment in (9), which has received quite a lot of recent philosophical attention. In fact, (9) underwrites a certain contentious argument for realism, known as THE EASY ARGUMENT FOR NUMBERS. Intuitively, (10) entails (6c).

(10)　Jupiter has four moons.

If so, and if (9) is a genuine entailment, where 'something' ranges over numbers,[6] then realism seemingly follows in virtue of successfully counting Jupiter's moons. As we will see, however, there is nothing approaching philosophical consensus as to whether (9) is a genuine entailment.

---

[5] I will return to this assumption in §3.3

[6] See Moltmann (2008) for relevant discussion.

Although I have framed this section around the Easy Argument, due to its prominence within the literature, it is important to appreciate its role within the context of Hale's Argument, from the outset. After all, apparent identity statements like (6c) are just one instance of number expressions purporting to function as singular terms within true numerical statements. Thus, even if the apparent entailment in (9) is not genuine, as some philosophers maintain, this alone would not undermine Singular Terms, given other examples like (6a,b). Rather, Singular Terms is false only if *no* terms purporting to refer to numbers are genuinely referential.

Put differently, whereas some philosophers have recently argued that the best available semantic evidence suggests rejecting the Easy Argument, this alone will not establish *non-referentialism*, consistent with nominalism. Conversely, if the only criteria for qualifying as a singular term is that an expression licenses entailments like (7-c), and yet the best available semantic evidence requires rejecting (9), then the Easy Argument would provide no support for realism. In this respect, the Easy Argument nicely illustrates the broader ontological question this monograph intends to address.

The rest of the section is laid out as follows. §2.1 discusses Frege (1884)'s observation that number expressions appear to have both referential and non-referential uses, and how this gives rise to the present philosophical debate. Specifically, it gives rise to two popular philosophical strategies—the substantival strategy and the adjectival strategy. The rest of the section then spells out the motivations for, and challenges facing, versions of these two strategies. Specifically, §2.2 discusses a version of the substantival strategy, suggested by Crispin Wright (1983), and the linguistic challenges for his neo-logicist program. §2.3 discusses three versions of the adjectival strategy. As we will see, while only one version purports to establish non-referentialism, all three face significant empirical challenges.

2.1 Two Strategies of Analysis

Gotlob Frege (1884, §57) notes that number words appear to have different uses in natural language. Specifically, they appear to have referential uses, as witnessed in (11b), and non-referential uses, as witnessed in (11a).

(11)   a. Jupiter has four moons.

        b. The number of Jupiter's moons is four.

One of Frege's major contentions is that while statements like (11a) appear to be about objects – e.g. Jupiter – they are really a statement about concepts – e.g. being a moon of Jupiter:

To throw light on the matter, it will help us to consider number in the context of a judgment that brings out its ordinary use. If, in looking at the same external phenomenon, I can say with equal truth 'This is a copse' and 'These are five trees,' or 'Here are four companies,' then what changes here is neither the individual nor the whole, the aggregate, but rather my terminology. But that is only a sign of the replacement of one concept by another. This suggests ... that a statement of number contains an assertion about a concept... If I say 'The King's carriage is drawn by four horses,' then I am ascribing the number four to the concept 'horse that draws the King's carriage.' (§46)

In modern logical notation, Frege contention can be rendered as (12), translating (11a), where 'M' abbreviates 'is a moon of Jupiter.'

$$(12) \quad \exists x_1, \ldots, x_4. \, [M(x_1) \wedge \ldots \wedge M(x_4) \wedge x_1 \neq x_2 \wedge \ldots \wedge x_3 \neq x_4] \wedge [\forall z. \, M(z) \rightarrow z = x_1 \vee \ldots \vee z = x_4]$$

The first conjunct asserts that at least four things are moons of Jupiter, while the second asserts that there are no other such moons. Together, then, (11a) will be true if Jupiter has exactly four moons.[7] On this analysis, 'four' is analyzed as a second-order property of concepts, namely those whose extensions consists of exactly four objects. Notice that (12) could be true in a model containing only five objects: four moons and Jupiter. Hence, the truth of (12) is consistent with nominalism.

Despite this, Frege goes on to observe that it is always possible to paraphrase apparently non-referential uses like (11a) in terms of apparent identity statements like (11b).

Since what concerns us here is to define a concept of number that is useful for science, we should not be put off by the attributive form in which number also appears in our everyday use of language. This can always be avoided. For example, the proposition 'Jupiter has four moons' can be converted into 'The number of Jupiter's moons is four.' Here the 'is' should not be taken as the mere copula ... Here 'is' has the sense of 'is equal to,' 'is the same as.' We thus have an equation that asserts that the expression 'The number of Jupiter's moons' designates the same object as the word 'four.' (§57)

Frege was primarily interested in developing an ideal logical language suitable for science: a *Begriffsschrift*. Within such a language, number expressions would always function referentially. Thus, Frege's suggestion is that non-referential uses like (11a) should be paraphrased as (11b), where the latter is an identity statement equating two numbers: the number of Jupiter's moons

---

[7] According to NASA (https://solarsystem.nasa.gov/moons/jupiter-moons/overview), Jupiter may have up to seventy-nine moons. Nevertheless, I will continue using Frege's examples due their influence, hoping no substantial confusion results.

and the number four. Formally, (11b) can be analyzed as (13), where '#' is a cardinality-function mapping a concept to a natural number (finite cardinal) representing how many objects fall under that concept.

(13) $\#[\lambda x.\, M(x)] = 4$

As an identity statement, (13) references a number directly. Thus, unlike (12), (13) would be true only if numbers exist. It would not be true in a model containing just four moons and Jupiter, for instance. So, whereas the truth of (12) is consistent with nominalism, (13) implies realism.

In sum, not only do number expressions appear to serve importantly different semantic functions, these different uses appear to have significantly different ontological consequences. Accordingly, Dummett (1991, p. 91) gives labels to two positions one might take in light of Frege's observation:

> Number-words occur in two forms: as adjectives, as in ascriptions of number, and as nouns, as in most number-theoretic propositions. When they function as nouns, they are singular terms, not admitting of the plural; Frege tacitly assumes that any sentence in which they occur as adjectives may be transformed either into an ascription of number ... or into a more complex sentence containing an ascription of number as a constituent part. Plainly, any analysis must display the connection between these two uses... Evidently, there are two strategies. We may first explain the adjectival use of number-words, and then explain the corresponding numerical terms by reference to it: this we may call *the adjectival strategy*. Or, conversely, we may explain the use of numerals as singular terms, and then explain the corresponding number-adjectives by reference to it; this we may call *the substantival strategy*.

Thus, according to THE SUBSTANTIVAL STRATEGY, apparently non-referential uses like (11a) are to be analyzed in terms of referential uses like (11b). In other words, (11a,b) both receive something like the ontologically committing truth-conditions in (13), following Frege. In contrast, according to THE ADJECTIVAL STRATEGY, apparently referential uses like (11b) are to be analyzed in terms of non-referential uses like (11a). Thus, both (11a,b) receive something like the ontologically non-committing truth-conditions in (12).

Obviously, Frege advocated the substantival strategy, at least for the purposes of developing an ideal logical language. However, despite Frege's considerable influence, we will see that the adjectival strategy has gained increasing popularity among philosophers in more recent times.

2.1.1 The Easy Argument for Numbers

Whether numbers exist is an important, longstanding philosophical question. Thus, it would be surprising if it could be answered simply by peering through a telescope and counting some moons. Yet that appears to be exactly what Frege's substantivalist analysis implies.

Indeed, if Frege's analysis is correct, then it would appear that we may validly infer that numbers exist on the basis of four seemingly plausible assumptions. The first is Equivalence:

Equivalence: (11a) and (11b) are truth-conditionally equivalent.

Indeed, (11a,b) are seemingly both true if Jupiter has exactly four moons. However, it is possible to deny Equivalence. After all, it may be argued that (11a,b) have different truth-conditions, based on contrasts like (14a,b).[8]

(14)   a. Jupiter has four moons. In fact, Jupiter has over seventy moons.

      b. ?? The number of Jupiter's moons is four. In fact, Jupiter has over seventy moons.

Whereas (14b) sounds explicitly contradictory, (14a) does not. Thus, (14a,b) purports to show that (11a) is true if Jupiter has *at least* four moons. However, this objection is easily avoided by inserting 'exactly' into (11a).

The second assumption required is Identity:

 Identity: (11b) is an identity statement.

This follows Frege's analysis, of course, given above in (13). Some motivation for Identity comes from comparing (11b) to sentences having similar syntactic structures. For example, consider (15), which according to Hofweber (2007), is a genuine identity statement.

(15)   The composer of *Tannhäuser* is Wagner.

So, if (15) is a genuine identity statement, then (11b) plausibly is too.

The next assumption follows immediately from Identity:

 Referentiality: 'four' in (11b) is a singular term.

Generally, identity statements take the form '$\alpha = \beta$,' where 'α' and 'β' are singular terms. So, if (11b) is a genuine identity statement, then 'four' in (11b) must be a singular term. The final assumption required is:

 Existential Generalization: Existential generalization is valid for singular terms.

In particular, if '$\alpha = \beta$' is true, then we may validly infer $\exists x. \alpha = x$, i.e. there is something which is α.

---

[8] See Horn (1972).

Now, suppose that (11a) is in fact true. Then (11b) is also true, by Equivalence. By Identity, (11b) is an identity statement having the form in (13). So, by Referentiality and Existential Generalization, (13) entails (16),

(16) $\exists n. \#[\lambda x. M(x)] = n \wedge n = 4$

paraphrased in English as (17).

(17)   There is a number which is the exact number of Jupiter's moons, namely four.

In sum, we may validly infer that numbers exist in virtue of successfully counting collections of objects.

This is puzzling. In the words of Hofweber (2007), (11a) appears to be "ontologically innocent," involving no commitment to numbers, whereas (11b) appears to be "ontologically loaded," seemingly involving explicit commitment to numbers. Even though (11a) and (11b) are seemingly equivalent, how is it that, to borrow a phrase from Schiffer (2003), we can get "something from nothing"? Is vindicating realism really that easy, and if so, why have realists and nominalists been arguing for so long? Understandably, then, this has become known as "the Easy Argument for Numbers."[9]

The Easy Argument is so-called because it seems far too easy. On the other hand, an argument's being relatively straightforward is not alone a good reason for *rejecting* it. What's needed, rather, are principled reasons for accepting or rejecting the premises, and both strategies purport to do just that. Specifically, if the substantival strategy is correct, then the Easy Argument would straightforwardly establish realism. But if the adjectival strategy is correct, then (11b) is not an identity statement after all, and so both (11a) and (11b) would be consistent with nominalism. In the remainder of the section, I will consider both strategies, their motivations, and their difficulties, beginning with substantivalism.

2.2 The Substantival Strategy

According to Frege's substantivalism, (11a) has the logical form of an identity statement, despite surface syntactic appearances. That is, despite 'four' in (11a) appearing to function as a non-referential expression, similar to 'large' or 'some' in (18), it is nevertheless analyzed as a numeral referring to a particular natural number.

(18)   Jupiter has {large/some} moons.

---

[9] See Balcerak-Jackson (2013), Snyder (2017), and other contributions in *Linguistics and Philosophy*, Vol. 40 (4), 2017.

Of course, Frege's analysis was designed for an ideal logical language, not natural language as such. However, at least one prominent philosopher has rhetorically suggested that this analysis may be actually correct for natural language. This section sketches its motivations and challenges.

2.2.1 Wright's Suggested Substantivalism

It is well known that the Dedekind-Peano axioms are derivable from HUME'S PRINCIPLE (HP) along with suitable additional definitions, a result known as FREGE'S THEOREM.[10] HP is an abstraction principle having the following form, where '$\approx$' designates the relation of EQUINUMEROSITY holding between two concepts just in case there is a bijection between their extensions.

(HP) $\forall F, G. \#[\lambda x. F(x)] = \#[\lambda x. G(x)] \leftrightarrow F \approx G$

In English, HP states that the number of Fs is identical to the number of Gs just in case the Fs and the Gs are equinumerous, i.e. each F can be mapped to a unique G, and vice versa.

The key contention of Hale and Wright (2001)'s neologicist program is that if HP can be independently justified, then since it implies the axioms characterizing the natural numbers, it can be seen as grounding basic arithmetic knowledge. We begin by stipulating HP as governing certain numerical identities, thereby defining a new concept of number. This stipulation is then justified in virtue of delivering the Dedekind-Peano axioms.[11] If so, and if it is also granted that anything following from definitions and logic alone is *analytic*, this would make the Dedekind-Peano axioms analytic, thus vindicating a restricted version of Frege's LOGICISM.[12]

Schematically, ABSTRACTION PRINCIPLES take the form in (19), where 'Σ' is a function mapping variables of type α and β to objects of that type, and '$\sim$' is an equivalence relation.

(19) $\forall \alpha, \beta. \Sigma(\alpha) = \Sigma(\beta) \leftrightarrow \alpha \sim \beta$

An example is Frege (1884)'s abstraction principle for directions, given in (20), where '$\mathcal{D}$' maps lines to directions, and '$\sim$' is the relation of parallelism.

(20) $\forall l, l'. \mathcal{D}(l) = \mathcal{D}(l') \leftrightarrow l \sim l'$

---

[10] See Heck (2011).

[11] See also Wright et al. (1999). Of course, this result alone will not establish HP as the uniquely *correct* characterization of the natural numbers, as there are numerous such characterizations available (see e.g. Snyder et al. (2018b)). Rather, this is the job of Frege's Constraint, discussed below.

[12] Broadly, logicism is the thesis that mathematics, or a subset of mathematics, is in some interesting sense derivable from logic and suitable definitions alone.

Accordingly, two directions are identical if the lines they are associated with are parallel. Speaking of (20), Wright (1983, p. 31–32) says:

> The reductionist idea was that since the right-hand contains no apparent direction-denoting singular term, we can take it that the apparent reference to a direction on the left-hand side is mere surface grammar, a misleading nuance. But why should we not turn that way of looking at things on its head? What is there to prevent us saying that, since the left-hand side does contain an expression referring to a direction, it is the apparent *lack* of reference to a direction on the right-hand side which is potentially misleading, or 'mere surface grammar'?…Why should it not be possible for a sentence containing no isolatable part which refers to a particular object nevertheless achieve, as a whole, a reference to that object—as is attested by the fact that it is equivalent to a sentence in which such a reference is *explicit*?

Wright's rhetorical suggestion appears to be that statements about parallelism like (21a), despite not containing any expressions which explicitly refer to directions, nevertheless manage to do so, in virtue of those statements being equivalent to identity statements which do contain expressions explicitly referring to directions, such as (21b).

(21) a. Line *l* and line *l′* are parallel.

      b. The direction of line *l* is identical to the direction of line *l′*.

Applying similar reasoning to HP, Wright's comments suggest that we might be able to refer to numbers via statements about equinumerosity, such as (22a), in virtue of being equivalent to statements which do explicitly reference numbers, such as (22b).

(22) a. The concept 'moon of Mars' and the concept 'moon of Haumea' are equinumerous.

      b. The number of Martian moons is identical to the number of Haumean moons.

And this suggests a natural route into the substantival strategy.

Specifically, although (11a) does not contain expressions which explicitly refer to numbers, it manages to do so anyway in virtue of being equivalent to a sentence which does explicitly reference a number, namely (11b). If so, then the correct analysis of (11a) is not one in which 'four' functions as a non-referential second-order concept, but rather as a singular term referring to a number, as in (13), thus revealing its actual logical form. We only need to recognize that the "mere surface grammar" of (11a) is misleading.

2.2.2 Challenges for Wright's Suggested Substantivalism

On Wright's rhetorical suggestion, despite surface syntactic appearances, 'four' in (11a) actually serves the same referential semantic function it does in (11b), thanks to their equivalence. There are two significant challenges for this suggestion, however. The first concerns whether the mere equivalence of (11a) and (11b) suffices to substantiate the substantival strategy. The second concerns the empirical legitimacy of Frege (1884)'s analysis of number expressions, and thus Wright's rhetorical suggestion. Specifically, it concerns the linguistic motivations for identifying the natural numbers with finite cardinal numbers. I will consider each in turn.

Dummett's Challenge

Does the purported fact that (11a,b) are equivalent license the conclusion that we may analyze 'four' in both examples as having the same referential semantic function, despite syntactic appearances? As Dummett (1991, p. 109) notes, the fact that equivalence is a symmetric relation suggests it does not. Criticizing Frege's proposed substantivalism, Dummett writes:

> If it is legitimate for analysis so to violate surface appearance as to find in sentences containing a number-adjective a disguised reference to a number considered as an object, it would necessarily be equally legitimate, if it were possible, to construe number-theoretic sentences as only appearing to contain singular terms for numbers, but as representable, under a correct analysis of their hidden underlying structure, by sentences in which number-words occurred adjectivally…If the appeal to surface form, in sentences of natural language, is not decisive, then it cannot be decisive, either, when applied to sentences of number theory. Frege has merely expressed a preference for the substantival strategy, and indicated a means of carrying it out.

The same can be said for Wright's rhetorical suggestion, of course. Assuming (11a) and (11b) are equivalent, if it is legitimate to ignore the surface syntax of the former and find reference to numbers where none seems to exist, then it should be equally legitimate to ignore the surface syntax of the latter and find quantification over objects where none seems to exist. Apparently, then, insisting that (11a) has the logical form of (13) amounts to merely expressing a preference for recognizing numbers as objects.

From a linguistic perspective, what we want, ideally, is a *compositional* explanation of why 'four' in (11a) functions referentially. In other words, how do the specific meanings of the component expressions involved combine so that the actual semantic representation of (11a) is (13), rather than (12)? Simply noting that (11a,b) are equivalent won't do *that*. Furthermore, there are *multiple* seemingly equivalent paraphrases of (11a) available, not all of which involve 'four' functioning referentially. Consider (23), for instance, where 'four' plausibly functions as a predicate.

(23)   Jupiter's moons are four (in number).

Why, then, shouldn't 'four' in all three cases be analyzed as a predicate instead? In short, the purported equivalence of (11a,b) is not sufficient to establish *either* of Dummett's strategies.

The Moltmann/Snyder Challenge

The second substantial challenge to Wright's suggested substantivalism comes indirectly through a wealth of data provided mostly by Friederike Moltmann (2013a,b). It purports to establish two key semantic theses. The first is that the noun 'number' is ambiguous between a monadic, arithmetic sense witnessed in (3a,b), and a relational, cardinal sense witnessed in (24a,b).

(3a)   Four is an even <u>number</u>.

(3b)   The <u>number</u> four is even.

(24)   a. John saw a <u>number</u> of women at the party.

      b. The <u>number</u> of women exceeded that of the men.

Secondly, it purports to establish that natural language draws a corresponding ontological distinction between *numbers* as arithmetic objects and *cardinalities* as representations of the cardinal size of collections. This, in turn, casts significant doubt on the (neo-)Fregean identification of numbers with cardinalities, and with it the empirical legitimacy of Wright's proposal.

Though Moltmann provides an abundance of evidence for these two claims, I will group that data into three kinds here for expository purposes. Most of it involves contrasts between what Moltmann calls EXPLICIT NUMBER-REFERRING TERMS like 'the number four' and THE NUMBER-OF TERMS like 'the number of Jupiter's moons.' According to Moltmann, though both sorts of expressions are usually referential, they refer to different sorts of objects, numbers and cardinalities, respectively. Snyder (2017) argues for a similar conclusion, based on similar sorts of observations.

The first kind of evidence involves contrasts with relational verbs such as 'notice' and 'count.' Whereas *the number of*-terms are generally acceptable with these, explicit number-referring terms are not.

(25)   a. Mary noticed the number {of women/??four}.

      b. Mary counted the number {of women/??four}.

      c. Mary compared the number {of women/??four} to the number of men.

      d. Mary was surprised by the number {of women/??four}.

Even if Mary happened to notice, count, compare, or be surprised by four women, it clearly does not follow that she noticed, counted, compared, or was surprised by *an abstract arithmetic object*. We see a similar contrast in predicates involving cardinal size.

(26)   a.   The number {of women/??four} is expanding rapidly.

   b.   The number {of women/??four} is larger than the number of men.

Again, if the cardinality of the women happens to be four, but is expanding rapidly, it clearly does not follow that an abstract arithmetic object is too.

   The second kind of evidence involves contrasts in apparent identity statements like (11b). Contrast (27a,b) for instance.

(27)   a.   The number of women is {four/??the number four}.

   b.   The number Mary is thinking about is {four/the number four}.

It is difficult to see why there would be such a contrast if *the number of*-terms referred to numbers, just like explicit number-referring terms such as 'the number Mary is thinking about.' In that case, we would expect 'the number four' to be acceptable in both examples, contrary to fact. Furthermore, we see the same contrast with 'how many'-questions.

(28)   a.   How many women came to the party? {Four/??The number four}.

   b.   Which number is Mary thinking about? {Four/The number four}.

Both pairs of contrasts make sense if English semantically distinguishes numbers and cardinalities as different *sorts* of entities. After all, it is commonly assumed that 'how many'-questions ask about cardinalities,[13] and it has also been argued that 'the number of women' in (27a) has a similar meaning.[14] If so, and if 'the number four' designates a number, then it is hardly surprising that the former is unacceptable in (27a,b).

   The final kind of data involves mathematical modifiers and functional relations. For example, only arithmetic 'number' is acceptable with mathematical modifiers such as 'natural,' 'rational,' and 'prime.'

(29)   a.   ?? the {natural/rational/prime} number of women

   b.   the {natural/rational/prime} number two

Similarly, only arithmetic 'number' is acceptable with functional expressions like 'successor of,' 'square root of,' and 'least common denominator of.'

---

[13] See e.g. Kennedy and McNally (2005) and Rett (2008).

[14] See Snyder (2017).

(30)　a.　?? the {successor/square root/least common denominator} of the number of women

　　　　b.　the {successor/square root/least common denominator} of the number four

Again, it is hard to see why this would be if *the number of*-terms and explicit-number referring terms were coreferential, both referring to numbers. It does make sense, however, if the former refer to cardinalities, the latter to numbers, and if natural language distinguishes these.

Suppose with Moltmann and Snyder then that natural language draws a semantic distinction between numbers and cardinalities. What would this imply for the (neo-)Fregean analysis? Most importantly, it would suggest that numbers are not *identified* with cardinalities in natural language. Thus, the numeral 'four' in e.g. (3a) and 'the number four' in (3b) are not coreferential with 'four' in (11b). Indeed, that the latter is intended to designate a *cardinality* is clear from the fact that '#' is paraphrased as 'the number of,' and also by non-equivalence of e.g. (31a,b).

(31)　a.　The number Mary is thinking about is four.

　　　　b.　$\#[\lambda x.\, \mathtt{Mary\text{-}thinking\text{-}about}(x)] = 4$

Obviously, Mary can think about the number four without thinking about four things. Again, this makes sense if (31a) is witness to monadic, arithmetic 'number,' the extension of which consists of numbers.

This appears to have significant consequences for the neo-logicist program. Again, the purpose of stipulating HP is to lay down truth-conditions for identity statements involving #-terms, thereby introducing a new concept of number. Now that we have distinguished monadic from relational 'number,' the question becomes: Which of these two senses does this novel concept introduce? And the answer, apparently, is: *cardinal* number. Thus, consider again Frege's analysis of (11b), given repeated in (13).

(13)　$\#[\lambda x.\, M(x)] = 4$

Again, 'number' in (11b) is clearly *relational*. Hence, '4' in (13) is only plausibly understood as referencing a *cardinality*, not a number. But then the truth of (11b) would not imply realism about *numbers*. Rather, at most it would entail realism about cardinalities.

To be clear, the problem is not that stipulating HP would fail to be justified. It would be, in virtue of implying the Dedekind-Peano axioms. That's because the finite cardinalities generated by HP are isomorphic to the naturals. Indeed, both form ω-sequences, and so conform to those axioms. The same holds for other potential characterizations of the naturals, such as that of Linnebo (2009), who shows that it is possible to derive the Dedekind-Peano axioms on the basis of a different abstraction principle, one which characterizes the naturals as *ordinals*, as opposed to cardinals.

Rather, the trouble is with another core neo-logicist commitment, known as FREGE'S CONSTRAINT.[15] Roughly, this states that a formal characterization of a set of numbers, such as the naturals, should directly reflect their empirical applications. To quote Frege (1903), those applications cannot be "tacked on externally."[16] To quote Wright (2000), they must be "absolutely on the surface." Simply put, if HP is to meet Frege's Constraint, then it needs to encode the empirical applications of the natural numbers without the help of additional deductive resources.[17]

Now, according Hale (2016), the primary empirical application of the naturals is what Benacerraf (1965) calls TRANSITIVE COUNTING, i.e. using numerals to answer 'how many'-questions. Moreover, it is assumed that HP satisfies Frege's Constraint given this application, while other candidate characterizations do not.[18] If so, then we have independent motivation for preferring HP as the uniquely correct characterization of the naturals.

Notice, however, that this argument goes through only if the referents of #-terms generated by HP are themselves assumed to be natural numbers. After all, if HP fails to generate naturals, then Frege's Constraint won't adjudicate between potential characterizations. Yet we have seen that HP more plausibly generates *cardinalities*, not numbers. Moreover, it would be hopeless to *subsequently* identify the naturals with finite cardinalities generated by HP on the basis of an observed isomorphism, since then applications would not be "absolutely on the surface." Rather, HP would be in a parallel predicament to other potential characterizations, such as Linnebo's.

2.3 The Adjectival Strategy

Despite the considerable philosophical influence of Frege's analysis, the adjectival strategy has gained increasing popularity in recent times. According to it, despite appearing to function as a numeral, 'four' in (11b) is said to have the same non-referential function witnessed in (11a). Consequently, 'four' in (11b) is not a genuine singular term.

---

[15] See Wright (2000) and Hale (2016).
[16] See Snyder and Shapiro (2016) for discussion.
[17] See Snyder et al. (2018a).
[18] However, see Snyder et al. (2018a, 2019) for arguments against this assumption.

Different versions of adjectivalism result depending on the purported semantic function of 'four' in (11a). Specifically, it may be a quantificational determiner like 'some,' or an adjective like 'large.'

(18)   Jupiter has {some/large} moons.

In either case, since identity statements equate the referents of genuine singular terms, Identity is false, thus rendering the Easy Argument unsound.

   Identity: (11b) is an identity statement.

In short, the adjectival strategy appears to provide a way of avoiding ontological commitment to numbers, at least with respect to Frege's original examples. Thus, nominalists may find this strategy attractive.

   However, to reemphasize, the adjectival strategy is not alone sufficient to undermine Singular Terms. Rather, only *non-referentialism* will do that, and adjectivalism, as defined by Dummett (1991), applies only to examples like (11a,b). Thus, one may instead opt to defend what Dummett calls THE RADICAL ADJECTIVAL STRATEGY: No single occurrence of a number word, including numerals like 'four' in (6a), is a singular term.

(6a)   <u>Four</u> is an even number.

However, this too will not suffice to undermine Singular Terms, in light of *complex* expressions like 'the number four.'

(6b)   <u>The number four</u> is even.

Even if 'four' in (6b) is not a singular term, phrases like 'the number four' or 'that number' purport to be. Thus, a complete defense of non-referentialism requires explaining why surface syntactic appearances are *radically* misleading with respect to semantic function, in this respect.

   As far as I'm aware, there has been only one articulated and sustained defense of non-referentialism within the philosophical literature, namely that of Thomas Hofweber (2005, 2007, 2016). As we will see, however, Hofweber's analysis suffers from numerous empirical problems, and alternative versions of adjectivalism have been subsequently developed which purport to overcome some of them. And while these alternatives face their own empirical challenges, the point I want to highlight here is this: they are not intended to vindicate non-referentialism in full generality. In fact, on the most comprehensive form of adjectivalism, namely Moltmann (2013a,b), complex expressions like 'the number four' in (6b) are singular terms, referring to numbers. Thus, even if the Easy Argument fails to establish realism, Hale's Argument potentially does, nevertheless.

2.3.1 Hofweber's Non-Referentialism

The version of non-referentialism defended by Hofweber (2005, 2007, 2016), which is largely responsible for sparking the present debate, is rather complex, consisting of numerous empirical hypotheses regarding numerical discourse and thought. Here I will highlight some of the core contentions.

Determiners, Extraction, and Focus Effects

The key semantic fact about natural language determiners, such as 'no,' 'some,' 'most,' 'every,' etc., is that they cannot function referentially. No empirically respectable semantics would claim that 'no,' for instance, can refer to an object. Rather, determiners combine with nouns like 'moon(s)' to form quantificational phrases such as 'no moons,' denoting second-order properties, or GENERALIZED QUANTIFIERS.

(32)   Jupiter has {no/some/four} moons.

On the predominant semantic theory of determiners, GENERALIZED QUANTIFIER THEORY (GQT; Barwise and Cooper (1981)), these uniformly denote relations between sets, as suggested in (33).

(33)   a. $[\![\text{no}]\!] = \lambda P.\lambda Q.\, P \cap Q = \varnothing$

      b. $[\![\text{some}]\!] = \lambda P.\lambda Q.\, P \cap Q \neq \varnothing$

      c. $[\![\text{four}]\!] = \lambda P.\lambda Q.\, |P \cap Q| = 4$

Here, 'P' and 'Q' range over predicate extensions (sets), and '$|\ |$' denotes the cardinality-operation on sets. Thus, 'no' denotes a relation between disjoint sets; 'some' denotes a relation between non-disjoint sets; and 'four' denotes a relation between sets whose intersection has exactly four members.[19]

The important observation is that determiners are not in the business of referring, semantically speaking. So, if 'four' in (11a) is a determiner, then it too must be functioning non-referentially. Hofweber's first key linguistic contention is that 'four' in (11b) is in fact *the very same* determiner in (11a), having something like the meaning in (33c).[20]

That's because, as Hofweber (2005, p. 211) explains, (11b) is actually the result of "moving" 'four' in (11a), via what Hofweber calls EXTRACTION.

> In Hofweber [2007], I argue that this focus effect can't be explained if one thinks that [(11b)] is both syntactically and semantically an identity statement with two (semantically) singular terms. But it can be explained if

---

[19] Actually, (33c), due to Breheny (2008), is a variation on the lower-bounded analysis in Barwise and Cooper (1981).

[20] Actually, Hofweber maintains ... Hofweber maintains that natural language determiners are "underspecified" for two "readings," an objectual "reading," corresponding to (33), and a substitutional "reading." See Hofweber (2016, Ch. 3) for details.

[(11b)] has a different syntactic structure, one that results from extracting the determiner and placing it in an unusual position that has a focus effect as a result. Thus, in [(11b)] 'four' is a determiner that has been "moved" out of its usual position.

The idea appears to be that through extraction, 'four' in (11a) gets "moved" to post-copular position, resulting in (11b). Despite this, 'four' retains its original semantic function as a non-referential determiner. To quote Hofweber (2005, p. 211): "The word 'four' is the same in [(11a)]) and [(11b)]." Thus, 'four' in both (11a,b) denotes a relation between sets, not a number.

Hofweber's main source of evidence for extraction concerns so-called FOCUS EFFECTS witnessed in examples like (34).

(34)   a. Johan likes soccer.

       b. What Johan likes is soccer.

       c. It is Johan who likes soccer.

Whereas (34a) is acceptable in response to both 'Who likes soccer?' and 'Which sport does Johan like?', (34b) is only acceptable in response to the latter, while (34c) is only acceptable in response to the former. Contrast this with prototypical identity statements like (35), which apparently do not give rise to focus effects.

(35)   Cicero is Tully.

Indeed, (35) is perfectly fine in response to both 'Who is Tully?' and 'Who is Cicero?'. In contrast, (11b) *does* display focus effects: while (11a) is acceptable in response to both 'Which planet has four moons?' and 'What belongs to Jupiter?', (11b) is only acceptable in response to the former. What this shows, according to Hofweber, is that (11b) is not a genuine identity statement. Moreover, it is indirect evidence that (11b) results from extraction since, if it were an identity, we would expect no focus effects.

Numerals and Semantically Bare Determiners

Since 'four' in (11b) is the same non-referential determiner witnessed in (11a), thanks to extraction, the truth of neither (11a) nor (11b) implies the existence of a number, no more so than (32) would.

(32)   Jupiter has {no/some} moons.

However, extraction is a construction-specific syntactic operation, presumably: it applies to sentences broadly having the structure of (11a), and returns sentences broadly having the structure of (11b). As such, it is not operative in arithmetic equations like (36), presumably.

(36)   Three plus two equals five.

After all, it is hard to see how (36) could result from anything similar to (11a), where the apparent numerals 'three,' 'two,' and 'five' feature originally as determiners. But then there is no obvious reason for thinking that the apparent numerals in (36) are non-referential.

. Thus, Hofweber distinguishes two kinds of BARE DETERMINERS, or determiners occurring without accompanying nouns, such as 'most' in (37).

(37)   How many boys kicked the ball? Most kicked the ball.

Although 'most' does not occur explicitly restricted by the noun 'boys' in (37), it is implicitly understood that way. In other words, the continuation of (37) is interpreted as 'Most boys…'. Contrast this with 'most' in (38), where no antecedent noun is available.

(38)   Most {is/are} more than none.

Rather than claiming something about most boys, or most people, or whatever, (38) is intended to be *generic*: whatever it is we're talking about, most is more than none. Hofweber calls determiners like 'most' in (38) SEMANTICALLY BARE DETERMINERS.

Hofweber's third key linguistic contention is that the number words in (36) are really semantically bare determiners, not genuine names of numbers. Put differently, (36) has something like the logical form informally suggested in (39), where 'X' is a noun phrase restricting the determiners 'three,' 'two,' and 'five,' and 'GEN' is a genericity operation.[21]

(39)   GEN: [three X and two (more) X are five X]

Hofweber's primary evidence for (39) is that arithmetic equations can be parsed in two ways, namely in the singular or in the plural, similar to (38).

(40)   Three and two {is/are} five.

Further, (40) also resembles (38) in that both are entirely general: no matter what we're talking about, three and two are five. Hence, despite surface syntactic appearances, (36) does not relate two first-order objects (3 and 2) to a third first-order object (5), through a first-order operation (+). Rather, it actually involves counting objects, though in an entirely general way.

This raises another question, however: What guarantees that things (the Xs) being counted in (39) do not overlap? This is crucial to getting the truth-conditions for (36) correct, of course: if $A = \{a, b, c\}$, $B = \{a, b\}$, and

---

[21]   See Krifka et al. (1995).

$C = \{d, e\}$, then $|A \cap B| = 2$, $|A \cup B| = 3$, and $|A \cup C| = 5$. So, what guarantees that (36) behaves like $|A \cup C|$, rather than $|A \cap B|$ or $|A \cup B|$? To this end, Hofweber appeals to a well-known distinction between PROPOSITIONAL (or BOOLEAN) CONJUNCTION and CUMULATIVE (or NON-BOOLEAN) CONJUNCTION. Examples of the former include (41a-c), while examples of the latter include (42a-c), inspired by Krifka (1990).

(41) a. John and Mary slept.

      b. Mary sang and danced.

      c. This cocktail is cheap and refreshing.

(42) a. John and Mary met at the mall.

      b. This concoction is beer and lemonade.

      c. That flag is entirely green and white.

(41a-c) can all be paraphrased as a conjunction of two propositions. For example, (41a) can be paraphrased as 'John slept and Mary slept.' In contrast (42a) cannot mean that John met at the mall, and also Mary met at the mall, just as (42b) cannot mean that this concoction is beer, and also this concoction is lemonade. What (41) and (42) show is that both kinds of conjunction can coordinate expressions having different semantic types – names, predicates, and modifiers.

Thus, Hofweber's fourth linguistic contention is that 'and' in (36) and (38) is cumulative conjunction involving semantically bare determiners, where non-overlap is guaranteed through "ellipsis, or a pragmatic mechanism, or a form of "free enrichment," or something else" (Hofweber (2005, p. 193)). Thus, Hofweber likens (36) to (43).

(43) She only had an apple and dessert.

Normally, an utterance of (43) would be judged misleading if she happened to have only an apple, though apples may be desserts. Presumably, according to Hofweber, this too is a function of "ellipsis, or a pragmatic mechanism, or a form of "free enrichment," or something else." The important point is that just as an utterance of (43) apparently presupposes non-overlapping extensions for 'apple' and 'dessert,' an utterance of (36) apparently presupposes non-overlapping extensions of 'three Xs' and 'two Xs.'

Arithmetic Reasoning and Cognitive Type-Coercion

An immediate consequence of Hofweber's semantics is that number expressions occurring within arithmetic statements should have a much more complex semantic type than meets the eye. Specifically, whereas names are generally assumed to have the simplest available extensional type (type e), determiners have a far more complex type (type $\langle\langle e, t\rangle, \langle\langle e, t\rangle, t\rangle\rangle$). Since 'and' generally coordinates expressions of the same type,[22] 'and' in 'Three and two is five' must thus have a *very* complex type: it takes two determiners as argument, and returns a determiner.

This appears to have significant cognitive ramifications. Specifically, basic arithmetic reasoning should seemingly be much more difficult than it really is. Hofweber (2005, p. 198-199) puts the point nicely:

> Thoughts that are expressed with bare determiners and that involve operations on determiners are quite unusual. In ordinary thinking, there are only very few cases of this besides the ones involving number determiners. 'Some but not all' and a few more come to mind, but their complexity is rather limited. Number determiners are special in this respect because they allow for the expression of complicated thoughts that involve essentially only bare number determiners and operations on them. Our minds, especially when we are young children, are not very well suited to reason with such thoughts.

Why, then, does arithmetic reasoning typically seem so easy if the meanings of the words involved are so complex?

To answer this, Hofweber appeals to THE LANGUAGE OF THOUGHT, arguing that our minds favor representations of objects over those of relations between sets. The resulting cognitive pressure leads to "coercing" the corresponding mental representations through an operation Hofweber (2005, p. 201) calls COGNITIVE TYPE-COERCION.

> We thus have a mismatch between the form of the representations that we want to reason with and the form of a representation that is required for our powerful reasoning mechanisms to be employed. But this mismatch can be overcome quite simply. We can force the representation to take on a form that fits our reasoning mechanism. The representation will have to change its form by systematically lowering the type of the bare determiners and operations on determiners to that of objects and operations on objects, respectively. Once this is done, the reasoning mechanisms we have can get a grip... I will call the process of changing the type of the form of a representation to facilitate cognition *cognitive type coercion*.

The basic idea appears to be that mental representations within the language of thought have something analogous to semantic types, which can be manipulated by cognitive operations analogous to type-shifting principles in natural language (more on this in Section 3). Specifically, when dealing with arithmetic statements, mental representations get coerced into representing numbers as objects, thereby facilitating arithmetic reasoning.

---

[22] Cf. Partee and Rooth (1983).

A significant consequence of cognitive type-coercion is that number words occupying argument positions within arithmetic statements retain their non-referential determiner meanings – hence why the kind of coercion involved is *cognitive*, rather than *semantic*. This suggests a potential non-referentialist strategy for dealing with apparent numerals such as 'two' in (3a).

(3a)   Two is an even number.

To quote Hofweber (2016, p. 145):

> Here there is a natural way to extend our account so far to cover cases like [(3a)]. After cognitive type coercion happens for bare determiner statements number determiners become available to be subjects in singular subject-predicate sentences. They still are not referential expressions on such uses, but determiners. Correspondingly, the predicate "is [an even] number" in [(3a)] is not predicated of the referent of "two," but forms a meaningful and true subject-predicate sentence nonetheless.

The idea seems to be that 'is an even number' is not a predicate true of numbers, but rather of relations between sets. The appearance of 'two' referring to a number is then presumably due to its occurring in argument position, thus resulting in cognitive type-coercion.

Internalism and the Simple Argument

We have seen that Hofweber's analysis supports non-referentialism, at least with respect to apparent numerals in sentences like (3a) and (11b). But what about other apparent singular terms, such as 'the number two'?

(3b)   The number two is even.

And what about (44), involving apparent quantification over numbers?

(44)   Some even number is prime.

Clearly, even if complex expressions like 'the number four' in (3b) failed to reference a number, the truth of (44) may support realism nevertheless.

To this end, Hofweber (2016, p. 107–108) argues for a thesis he calls INTERNALISM about a particular domain of things:

> Internalism about a domain of discourse is the view that in general the [apparent] singular terms in that domain are not referential, and that the quantifiers are in general used internally.

To spell this out, we first need another thesis Hofweber adopts concerning natural language determiners, namely that all natural language determiners are semantically "underspecified" for two "readings": an INFERENTIAL, or INTERNAL reading, and a DOMAIN CONDITIONS, or EXTERNAL reading. These two "readings" correspond respectively to a well-known distinction between SUB-STITUTIONAL and OBJECTUAL QUANTIFICATION. Roughly, according to the latter, 'Something is F' is true in a model if the domain contains at least one object satisfying F, while according to the former, it is instead true in a model if there is at least one term t in the language such that '$F(t)$' is true in that model. Thus, generally speaking, quantification in natural language can be understood either substitutionally or objectually, according to Hofweber.

Internalism about F-discourse is then the conjunctive thesis that i) non-referentialism about terms purporting to refer to Fs is correct, and ii) quantificational talk about Fs is generally substitutional. Specifically, internalism about natural number discourse maintains that terms purporting to refer to natural numbers never actually do so, and quantification over natural numbers is generally substitutional.

Suppose so. Then by non-referentialism, 'the number four' in (3b) is not a genuine singular term, and 'some' in (44) is interpreted substitutionally: there is a term, e.g. 'two,' substitutable for the relevant predicates. Hofweber's internalism thus not only renders numerical discourse consistent with nominalism, according to Hofweber (2016, p. 111), it actually implies that nominalism with respect to the natural numbers is *correct*, thanks to what he calls THE SIMPLE ARGUMENT.

> If internalism is true about talk about natural numbers, then number words are non-referring expressions. Thus expressions like "2" or "the number 2" do not refer, in the broad sense, to anything. Thus none of the objects in the domain of our external quantifiers are referred to by these expressions. And thus none of them is the number 2. Since "the number 2" does not pick out or denote any object, whatever objects there may be, none of them is the number 2. So among all the objects, none of them is the number 2.

Let 'd' name an object in the domain of our natural language quantifiers. Since identity statements equate the referents of genuine singular terms, no statement of the form 'd is the number n' is true. So, d is not the number one, or the number two, ... Thus, given the equivalence in (45),

(45) d is a natural number iff d is the number one or d is the number two or ...

it follows that d is not a natural number. Generalizing, no object in the domain of our natural language quantifiers is a natural number. Hence, (46) is *objectually* true.

(46)   Nothing is a natural number.

And this is just a restatement of nominalism with respect to natural numbers: no objects are natural numbers.

Summarizing, Hofweber's analysis may be viewed as a sustained defense of nominalism with respect to the natural numbers.[23] Contrary to surface appearances, apparent singular terms referring to those numbers are just that: apparent. Moreover, quantification over natural numbers is, despite appearances, similarly non-committing. Ultimately, then, not only is the linguistic evidence consistent with nominalism about the natural numbers, according to Hofweber, it in fact *recommends* it.

Challenges for Hofweber's Non-Referentialism

Hofweber's analysis consists of several broadly empirical theses, including:

(HT1) 'four' in (11b) is the same non-referential determiner witnessed in (11a), thanks to extraction.

(HT2) The kind of "movement" involved in extraction creates a focus effect, unobserved in prototypical identity statements.

(HT3) Apparent numerals occurring in arithmetic equations are actually semantically bare determiners.

(HT4) As a result, arithmetic statements involve expressions with complex semantic types. This creates a kind of "mismatch" in mental representations, thus triggering cognitive type-coercion, wherein numbers are represented as objects.

(HT5) Thanks to cognitive type-coercion, apparent numerals figuring in subject-predicate sentences appear to function as genuine singular terms.

(HT6) All natural language determiners are "underspecified" for two "readings," corresponding to objectual and substitutional quantification.

(HT7) Internalism about natural number discourse is correct: no terms apparently referring to natural numbers are genuinely referential, and quantificational talk about natural numbers is generally substitutional.

(HT8) Internalism about natural number discourse entails nominalism with respect to the natural numbers.

Each of these claims is highly contentious, and most are of dubious empirical legitimacy. Indeed, despite its influence, Hofweber's non-referentialism has received a fair amount of criticism in the literature. This has focused largely on (HT1) and (HT2). Here, I want to briefly look at those objections, while also casting doubt on Hofweber's other theses.

---

[23] Officially, Hofweber is neutral with respect to whether other abstracta exist, e.g. real numbers or sets. See Shapiro et al. (ms) for relevant discussion.

Let's begin with extraction. It is natural to interpret Hofweber's talk of "movement" as an instance of the same kind of "movement" familiar from transformational syntax. For example, consider (47b), where the blank indicates the position out of which "movement" of the underlined expression in (47a) is assumed to occur.

(47) a. Something that we weren't expecting happened.

b. Something happened that we weren't expecting.

As a result, 'that we weren't expecting' is said to become *focused*, much like post-copular 'four' in (11b) on Hofweber's analysis. It should thus be unsurprising that some of Hofweber's detractors, chiefly Balcerak-Jackson (2013) and Moltmann (2013b), have interpreted extraction as a transformational mechanism responsible for "rearranging" the syntactic material in (11a).

The problem, according to these detractors, is that the actual principles or operations required to do this kind of "rearranging" would not be recognized by contemporary transformational theories, and their postulation would be highly dubious. For one thing, unlike with (47b), (11b) clearly contains material missing in (11a): 'the,' 'number,' 'of,' and '-'s.' Conversely, there is material contained in (11a) that is missing in (11b): "has." Even if there were "movement" of the parts of (11a), no other known transformational mechanism would delete and add material in the manner required. Yet without (11b) resulting from some kind of "rearrangement" of (11a), there would be no guarantee that post-copular 'four' in (11b) is the *same* non-referential determiner (11a), thereby vindicating (HT1).[24]

A different criticism concerns (HT2). Hofweber's claim is that because genuine identity statements do not exhibit focus effects, but (11b) does, (11b) cannot be a genuine identity statement. However, Brogaard (2007) points out that (15), which Hofweber (2007) claims to be a genuine identity statement, exhibits similar focus effects.

(15) The composer of *Tännhäuser* is Wagner.

In particular, (15) would be an appropriate answer to the question 'Who composed *Tännhauser*?' but not 'Who is Wagner?' or 'What did Wagner do?' Thus, it appears that exhibiting focus effects is insufficient to warrant Hofweber's intended conclusion, namely that (11b) is not an identity statement, even if 'four' in (11b) results from extraction.

There are also potential empirical problems with (HT3) and (HT5). Specifically, it appears to make numerous false semantic predictions, as revealed

---

[24] For a response to this objection, see Hofweber (2014). And for criticisms of this response, see Snyder et al. (ms.).

by a number of semantic contrasts due to Rothstein (2017). First, 'count' is ambiguous between two senses, roughly corresponding to Benacerraf (1965)'s distinction between transitive and INTRANSITIVE COUNTING (listing numerals in order), witnessed respectively in Rothstein's (48b,a).

(48)   a. I counted to thirteen (??things/??people/??books).

     b. I counted thirteen (things/people/books).

As the labels suggest, transitive 'count' requires a direct object, where intransitive 'count' does not. Semantically, this suggests that while transitive 'count' has an essentially relational meaning, intransitive 'count' does not. Thus, consider Rothstein's (49a,b):

(49)   a. I counted thirteen. – Thirteen what?

     b. ?? I counted to thirteen. – Thirteen what?

It's hard to see how there could be a difference in acceptability in (48) and (49), however, if 'thirteen' functions uniformly as a semantically bare determiner in these examples. Secondly, (50a,b) are clearly not synonymous, as (50b) is clearly false.

(50)   a. Two is an even prime.

     b. Two things are even primes.

Yet if 'two' in (50a) were a semantically bare determiner, then (50a) would be synonymous with (50b), contrary to fact. Finally, numerals and bare determiners differ in their agreement features. Specifically, whereas numerals require singular morphology, bare determiners require plural morphology.

(51)   a. Which one of these three numbers is Mary's favorite? Four {is/??are}.

     b. How many people are coming to the party? Four {??is/are}.

Again, this would not be expected if the numeral 'four' in (51a) were a semantically bare determiner.

In short, the linguistic evidence strongly suggests that numerals cannot generally be semantically bare determiners. Rothstein (2017, p. 28) thus reasonably concludes: "Together, these data show that ... there are cases where a bare cardinal numerical must be a singular term." If so, then (HT3) is false, and the appearance of referentiality is not to be explained via cognitive type-coercion, contra (HT5).

Finally, there are good reasons to doubt that Hofweber's other theses actually support nominalism about the natural numbers. For example, suppose (HT4) is correct. Then within Mentalese, numbers are represented as objects. If so, then there would appear to be an analog of Hale's Argument available. Specifically, if numbers are represented as objects within the language of thought, then why shouldn't our ability to think and reason successfully about them be just as good a reason to think that they exist as our apparent ability to *speak* truthfully about them? Why think that only *natural* language is ontologically committing?

Now consider (HT6)-(HT8). One can, and should, question the empirical and theoretical motivations for these claims. First, note that (46) supports nominalism only if it is *objectually* true, i.e. no natural numbers are in the domain of our natural language quantifiers.

(46)   Nothing is a natural number.

However, this appears to be inconsistent with the internalist thesis that quantificational talk about natural numbers is generally *substitutional*. Clearly, (46) is substitutionally false, since there are terms substitutable for the relevant predicate, e.g. 'two.' On the other hand, if quantificational number talk is sometimes objectual, as (46) requires, then nothing obviously prevents (44) from being objectually true.

(44)   Some even number is prime.

In that case, however, some object would be a (natural) number, thus undermining Hofweber's intended nominalism. In short, even if (HT6) were correct, (HT7) and (HT8) may not be jointly consistent.

2.3.2 Felka's Adjectivalism

We have seen that much of the extant criticism leveled at Hofweber's Non-Referentialism is directed at "extraction." However, the next version of adjectivalism considered here, due to Katharina Felka (2014), avoids these objections. According to Felka, (11b) is not the result of "extraction," but a different syntactic operation: ELLIPSIS. It should be emphasized, however, that because Felka's adjectivalism is only intended to apply to apparent identity statements like (11b), even if that analysis were successful, it would not alone undermine Singular Terms.

Equative, Predicational, and Specificational Sentences

Felka's analysis presupposes a well-known distinction between three kinds of copular sentences. These are:

(52)   a. Cicero is Tully.   (EQUATIVE)

   b. Cicero is bald.   (PREDICATIONAL)

   c. The most famous Roman orator is Tully.   (SPECIFICATIONAL)

The labels 'equative,' 'specificational,' and 'predicational' are due to Higgins (1973).[25] Equative sentences are prototypical identity statements like (52a), equating the referents of two singular terms. Predicational sentences such as (52b) predicate a property like being bald of the subject. Finally, specificational sentences such as (52c) specify an individual under a certain description, e.g. the most famous Roman orator.

There are good linguistic reasons for thinking that specificational sentences and equatives are semantically distinct. For example, Mikkelsen (2005) points out that the gender-neutral pronoun 'it' is acceptable in tag-questions following specificational sentences but not predicational sentences.

(53)   a. Cicero is Tully, isn't {he/??it}?

        b. Cicero is bald, isn't {he/??it}?

        c. The most famous Roman orator is Tully, isn't {he/it}?

(53c) reveals that specificational sentences differ from equatives in this respect as well. That's plausibly because equative sentences are a special case of predicational sentences. In particular, equatives predicate the property of being identical to the individual referenced in the predicate.[26]

Mikkelsen also observes a similar contrast in overt questions like (54a,b).

(54)   a. Q: What is the most famous Roman orator?
           A: {He/??That/??It}'s bald.

        b. Q: Who is the most famous Roman orator?
           A: {He/That/It}'s Tully.

The question in (54a) has the form of a predicational sentence – it asks about a general property of an individual under a certain description – while the question in (54b) has the form of a specificational sentence – it asks about the identity of an individual under a certain description. As Mikkelsen explains, it is reasonable to expect anaphoric pronouns like 'he,' 'that,' and 'it' to track certain referential features of their antecedent subjects, e.g. their being a person. What (53) and (54) reveal is that predicational sentences do this but specificational sentences do not. Consequently, Mikkelsen reasonably concludes that whereas the subjects of predicational sentences are singular terms, those of specificational sentences are not. If so, then specificational sentences cannot generally be synonymous with equative sentences, assuming 'Cicero' in (52a) is a singular term.

---

[25] Higgins originally distinguished four kinds of copular clauses. However, the three-way taxonomy assumed here follows Mikkelsen (2005, 2011).

[26] See Partee (1986a).

The Question-in-Disguise Analysis

For present purposes, the significance of Higgins's taxonomy is that (11b) appears to be more plausibly viewed as a specificational sentence than an equative sentence. In fact, as Moltmann (2013b) points out, one of Higgins's original examples of a specificational sentence was (55), which is virtually identical in structure to (11b).

(55)　The number of planets is eight.

If so, then this suggests that the problem with the Fregean analysis is that it conflates specificational sentences with genuine identity statements. Nevertheless, this alone will not block the entailment from (11b) to realism. After all, even if (11b) is not an identity statement, it still could be that post-copular 'four' functions as a singular term, referring to a number.[27] In that case, even though Frege's substantivalist analysis would be flawed, it still would not follow that an adjectivalist analysis is correct.

　Felka's adjectivalism blocks this potential rejoinder by adopting a particular analysis of specificational sentences, namely THE QUESTION-IN-DISGUISE ANALYSIS (QDA).[28] According to QDA, specificational sentences express question-answer pairs, where the pre-copular material expresses a question, an answer to which is expressed in the post-copular material, both via ellipsis. For example, (52b) is said to have the underlying syntax in (56), where strike-through represents ellipsis.

(56)　[~~Who~~ the most famous Roman orator ~~is~~] is [Tully ~~is the most famous Roman orator~~]

According to (56), the pre-copular definite in (52b) results from eliding an indirect version of the question 'Who is the most famous Roman orator?', while the post-copular name 'Tully' results from eliding a complete sentence answering that question – 'Tully is the most famous Roman orator.'

　Felka proposes a similar analysis for (11b), given in (57).

(57)　[~~What~~ the number of Jupiter's moons ~~is~~] is [~~Jupiter has~~ four ~~moons~~].

So, the pre-copular material in (11b) actually expresses an indirect version of the question 'What is the number of Jupiter's moons?', while the post-copular material expresses an answer to that question, namely (11a). Since 'four' in (11a) is a quantificational determiner, according to Felka, post-copular 'four' in (11b) is non-referential. Thus, we can get the same effect of Hofweber's extraction by appealing to a less contentious syntactic operation, namely ellipsis. Call this THE ELLIPTICAL ANALYSIS.

---

[27]　See §2.3.3 and §3.3.3.

[28]　See Schlenker (2003).

Challenges for Felka's Adjectivalism

Despite its apparent linguistic advantages, the Elliptical Analysis faces some challenges. Some of these resemble those facing Hofweber's analysis. For example, because 'four' in (11b) is a determiner resulting from eliding most of the material in (11a), it stands to reason that a similar syntactic operation should be available for other determiners, e.g. those in (32).

(32)   Jupiter has {no/some} moons.

Specifically, it seems that the following analysis ought to be available.

(58)   [~~What~~ the number of Jupiter's moons ~~is~~] is [~~Jupiter has~~ {no/some} ~~moons~~].

After all, both (59a,b) seem like perfectly good responses to the question 'What's the number of Jupiter's moons?'.

(59)   a.   Jupiter has no moons.

        b.   Jupiter has some moons (though I don't know exactly how many).

Yet, seemingly as with Hofweber's extraction,[29] the sentence that would result from (58) via ellipsis is clearly unacceptable.

(60)   ?? The number of Jupiter's moons is {no/some}.

Thus, the question facing both analyses is this: Why can number words, but *no uncontroversial determiners*, occupy the post-copular position of specificational sentences?

Another important question for the Elliptical Analysis has to do with Felka's particular implementation of QDA. Greenberg (1977) originally observes, and Heim (1979) and Frana (2006) later report, a semantic distinction in so-called CONCEALED QUESTIONS like (61a,b).[30]

(61)   a.   John found out the murderer of Smith.

        b.   John found out who the murderer of Smith was.

Intuitively, whereas (61a) would be true only if John found out the exact identity of Smith's murderer, (61b) could be true even if John found out some less specific fact about Smith's murderer. Suppose, for instance, that all John discovered is that Smith's murderer, *whoever that was*, previously went by the name 'Jerry' and sold used cars in Minneapolis. Thus, (61a) and (61b) are not synonymous, and so 'the murderer of Smith' cannot mean 'who the murderer of Smith was.'

---

[29]   Cf. Balcerak-Jackson (2013).

[30]   For a discussion of concealed questions, see Nathan (2006).

Now, according to Felka's analysis, the pre-copular material in (11b) expresses an indirect version of the question 'What is the number of Jupiter's moons?'. Moltmann (2013b) instead suggests that it expresses an indirect version of 'How many moons does Jupiter have?' The data in (62) appear to support Moltmann's suggestion.

(62) a. John knows the number of Jupiter's moons, namely {four/??even}.

  b. John knows what the number of Jupiter's moons is, namely {four/even}.

  c. John knows how many moons Jupiter has, namely {four/??even}.

After all, 'the number of Jupiter's moons' in (62a) patterns like 'how many moons Jupiter has' in (62c), not 'what the number of Jupiter's moons is' in (62b). This makes sense if 'John knows the number of Jupiter's moons' and 'John knows how many moons Jupiter has' both require John to know the exact numerical identity of Jupiter's moons, while 'John knows what the number of Jupiter's moons is' only requires John to know some less specific fact about their cardinality, e.g. it's being large or even. Thus, it appears (62a,b) are not synonymous, and so the pre-copular material cannot express 'what the number of Jupiter's moons is,' contra Felka's prediction.

2.3.3 Moltmann's Radical Adjectivalism

Like Felka, Moltmann (2013a,b, 2017) argues that (11b) is a specificational sentence. However, whereas Felka's analysis is designed for specificational sentences in particular, Moltmann's analysis accounts for a broader range of data, including bare numerals occurring in various arithmetic contexts. Ultimately, Moltmann argues that the natural language evidence best supports radical adjectivalism. However, because English contains explicit number-referring terms such as 'the number two,' which in fact refer to numbers, the truth of examples like (3b) nevertheless supports realism.

(3b) The number four is even.

Because Moltmann's analysis is fairly complex, it will help to distinguish a number of empirical theses it is designed to defend.

(MT1) The English noun 'number' is ambiguous between a monadic, arithmetic sense, and a relational, cardinality sense.

(MT2) Natural language draws a corresponding sortal distinction between numbers qua abstract arithmetic objects and cardinalities qua concrete representations of how many individuals belong to a plurality.

(MT3) Explicit number-referring terms such as 'the number four' refer to pure numbers, i.e. abstract arithmetic objects. However, in such terms, 'number' is not a genuine sortal predicate, but rather serves a kind of REIFICATION function.

(MT4) *The number of*-terms refer instead to cardinalities. Cardinalities are NUMBER TROPES, i.e. concrete, particularized number properties, such as being four in number, instantiated in a plurality.

(MT5) Pluralities are not "set-like" entities existing in their own right, as on the model of singular reference assumed within linguistic semantics (see Section 3). Rather, they are individuals referenced "plurally," as on the model of PLURAL REFERENCE assumed within plural logic.[31]

(MT6) (11b) is a specificational sentence, with the pre-copular material expressing the question 'How many moons does Jupiter have?', and the post-copular material expressing (11a).

(MT7) Bare numerals are QUASI-REFERENTIAL EXPRESSIONS: they have the syntactic status of a full noun phrase, but retain their non-referential semantic function as cardinal adjectives.[32]

The main motivations for (MT1)–(MT4) were discussed in §2.2.2. For example, the thesis that cardinalities are best understood as number tropes is supported by the fact that they can combine with predicates of perception.[33]

(25a)   Mary noticed the number {of women/??four}.

The reifying function of 'number' in explicit number-referring terms is defended in Moltmann (2013a); (MT5) is defended in Moltmann (2016); and the main motivations for (MT6) were discussed in the previous section.

I want to focus here instead on the data motivating (MT7),[34] for two reasons. First, if correct, it would eliminate the substantival strategy as a genuine possibility. Secondly, it would appear to undermine what one might take to be a kind of best case "easy argument" for realism. In particular, (63b) seems like a genuine entailment of (63a), and the former appears to explicitly acknowledge the existence of a number.

(63)   a. Four is even.

b. There is a number which is even, namely four.

---

[31] For a discussion of singular and plural reference and their applications to semantics, see Carrara et al. (2016).

[32] See Moltmann (2017).

[33] See Moltmann (2013a) for an extended discussion of tropes and their uses in natural language semantics.

[34] Moltmann (2013b) also offers syntactic arguments for (MT7). However, since these are retracted in Moltmann (2017), I will not consider them here.

Similarly, (64b) seems like a genuine entailment of (64a).

(64)   a. Four added to three equals seven.

       b. There is a number, namely four, which when added to three equals seven.

However, if 'four' in (63a) and (64a) is actually a numeral functioning as a non-referential adjective, as required by (MT7), then neither entailment is licensed, thus blocking the intended realist conclusion.

Evidence for Quasi-Referentiality

Moltmann provides two kinds of evidence for the quasi-referentiality of numerals. The first involves numerals occurring in non-mathematical environments such as 'twelve' in contrasts like (65) and (66) (all judgments reported in this section are Moltmann's).

(65)   {The number twelve/??Twelve}, which interests me a lot, plays an important role in religious and cultural contexts.

(66)   a. Twelve is a number I would like to write my dissertation about.

       b. ??? I would like to write my dissertation about twelve.

It is hard to see how 'twelve' in (65) could be unacceptable if it were a genuine singular term, one coreferential with 'the number twelve.' Similarly, it is hard to see how there could be any contrast in (66a,b) if 'twelve' were functioning as a genuinely referential name in both examples. At the same time, 'twelve' is clearly functioning as a full noun phrase in both (65) and (66a,b), i.e. it is in fact a numeral. Thus, these contrasts purport to reveal that 'twelve' in (65) and (66a) is not functioning as a singular term, but rather as a *quasi*-referential expression.

   The second kind of evidence involves bare numerals occurring in arithmetic environments. As Hofweber (2005) observes, there are different ways of parsing equations like '3 + 2 = 5,' using singular or plural morphology.

(40)   Three and two {is/are} five.

If these expressed genuine identity statements, then the terms flanking the copula could be swapped, similar to e.g. 'Cicero is Tully' and 'Tully is Cicero.' However, Moltmann contends that this is not possible.

(67)   ?? Five is three and two.

Furthermore, explicit number-referring terms are not acceptably substitutable for the numerals in (40).

(68)   ?? The number three and the number two {is/are} the number five.

Again, this is not what we'd expect if the numerals in (40) were genuinely referential. Next, Moltmann notes that '3 + 2 = 5' can be similarly parsed in the singular or plural with 'make,' and we see the same contrast with explicit number-referring terms.

(69)   a. Three and two {makes/make} five.

        b. ?? The number three and the number two {makes/make} the number five.

And, again, the terms flanking the verb cannot be swapped.

(70)   ?? Five makes three and two.

Finally, whereas explicit number-referring terms are acceptable with identity predicates like 'the same number as,' bare numerals apparently are not.

(71)   a. The number five is the same number as the number five.

        b. ?? Three and two {is/are} the same number as five.

        c. ?? Three and two {make/makes} the same number as five.

        d. ?? Five is the same number as three and two.

According to Moltmann, all of this suggests two conclusions. First, despite paraphrasing numerical identities, (40) is not a genuine identity statement. Secondly, and more generally, bare numerals are quasi-referential.

Moltmann's Analysis of Basic Arithmetic Statements

Given the previous data, Moltmann proposes that (40) and (69a) have a meaning roughly paraphraseable as (72).

(72)   If there were three things and two more things, then there would be five things.

Spelling this out, Moltmann analyzes cardinal adjectives as predicates true of pluralities consisting of a certain number of individuals. For example, 'four' receives the denotation in (73), where 'xx' is a plural variable and '$\leq$' is the primitive relation of INCLUSION, pronounced 'is one of' or 'is/are among,' holding between individuals.[35]

(73) $[\![four]\!] = \lambda xx. \exists y_1, \ldots y_4.[y_1 \leq xx \ldots \wedge y_4 \leq xx \wedge y_1 \neq y_2 \ldots \wedge y_3 \neq y_4]$
        $\wedge [\forall z. z \leq xx \rightarrow z = y_1 \vee \ldots z = y_4]$

---

[35] For an overview of plural logic and its use of inclusion to avoid talk of set-like entities, see e.g. Boolos (1985) and Oliver and Smiley (2013).

Thus, 'four' is a (first-order) predicate true of some individuals xx just in case exactly four individuals are among xx. Next, Moltmann analyzes 'and' and 'is' in (40) syncategorematically, as what she calls THE IS OF CALCULATION, synonymous with 'make' in (69a). In other words, 'and' and 'is' in (40) both have special, unanalyzable meanings when occurring together in arithmetic contexts, as do 'and' and 'make' in (69a), though both combinations are synonymous. Ultimately, combining these with cardinal denotations leads to the following truth-conditions for (40) and (69a).

(74) $\Box[(\exists xx.\exists yy.\ \texttt{three}(xx) \land \texttt{two}(yy) \land \neg\exists z.\ z \leq xx \land z \leq yy) \rightarrow$
  $(\exists zz.\ \texttt{five}(zz) \land xx \leq zz \land yy \leq zz \land \neg\exists z.\ z \leq zz \land \neg z \leq xx \land \neg z \leq yy)]$

In English, (74) states that, necessarily, if there are three things and two more things, then there are a total of five things.[36]

But if bare numerals are always non-referential cardinal adjectives, how to explain the apparent entailments between (63a,b) and (64a,b)?

(63a) Four is even.

(63b) There is a number which is even, namely four.

(64a) Four added to three equals seven.

(64b) There is a number, namely four, which when added to three equals seven.

According to Moltmann (2017), (75a) is substitutable for (63a), and (75b) for (64a), thanks to the reifying function of arithmetic 'number.'

(75) a. The number four is even.

  b. The number four added to three equals seven.

Since 'the number four' refers to a number, (75a) entails (63b), just as (75b) entails (64b). Thus, even though (63a,b) and (64a,b) are not genuine entailments, the fact that we can substitute explicit number-referring terms for their bare numeral counterparts might explain why they appear to be.

Challenges for Moltmann's Radical Adjectivalism

Although Moltmann's analysis is arguably the most empirically well-motivated version of adjectivalism available, it still faces some challenges. The first concerns (MT6), the thesis that (11b) is a specificational sentence, with the pre-copular material expressing the question 'How many moons does Jupiter

---

[36] This is very similar to an analysis proposed originally by Hodes (1984), though implemented within plural logic.

have?', and the post-copular material expressing (11a) as an answer to that question. Moltmann (2013b) cites two importantly different kinds of analyses of specificational sentences, though without officially committing to either. One appeals to ellipsis, namely QDA. However, this clearly won't suffice to derive a 'how many'-question as the pre-copular material of (11b), since no amount of eliding 'how many moons Jupiter has' will result in 'the number of Jupiter's moons.' This is presumably why Felka analyzes the pre-copular material as 'what the number of Jupiter's moons is' instead.

The other kind of analysis, due to Romero (2005), provides a semantic characterization of specificational sentences. According to it, the specificational copula has the denotation in (76), where "y" ranges over INDIVIDUAL CONCEPTS, or functions from worlds to individuals.

(76) $\llbracket \text{be} \rrbracket = \lambda x.\lambda y_{\underline{<s,e>}}.\lambda w.\, \underline{y}(w) = x$

According to this INDIVIDUAL CONCEPT ANALYSIS (ICA), the pre-copular material of a specificational sentence denotes an individual concept, while the post-copular material functions as a singular term. Thus, 'the number of Jupiter's moons' in (11b) would denote a function from worlds to cardinalities representing how many moons Jupiter has in that world, and post-copular 'four' would presumably function as a name referring to a specific cardinality. Thus, the trouble is that if ICA is correct, then post-copular 'four' must be functioning *referentially*. In other words, ICA is apparently inconsistent with (MT7). In either case, then, it is not clear how (11b) can come to express the 'how many'-question required by (MT6).

Another important question for Moltmann concerns the data motivating (MT7), much of which is suspect. For example, with some prior context, (65) and (66b) are less obviously contrastive, and certainly not *impossible*, as Moltmann's judgments indicate.

> *Context*: John asks: "I overheard you talking about a certain number you were considering writing your thesis about. Which number was that again?". Mary responds:

(65)   {The number twelve/Twelve}, which interests me a lot, plays an important role in religious and cultural contexts.

(66b)   I would like to write my dissertation about twelve.

Likewise for certain data paraphrasing arithmetic equations, such as (67), which is also less clearly unacceptable with prior context.

Context: A number of arithmetic sums are on the blackboard, e.g. '1 + 6' and '3 + 2.' John asks Mary: "Which number is three and two?". Mary responds:

(67)   Five is three and two.

Now, recall the explanation for the apparent entailment from (63a) to (63b) proposed above, according to which the genuinely referential 'the number four' is substitutable for non-referential 'four.'

(63a)   Four is even.

(63b)   There is a number which is even, namely four.

If this explains why (63a) apparently entails (63b), then shouldn't a similar explanation be available for the apparent entailment from (65) to (77)?

(77)   There is a number which interests me a lot and plays an important role in religious and cultural contexts, namely twelve.

If so, then shouldn't (65) with 'twelve' be *equally* acceptable, given that (65) with 'the number twelve' is, by hypothesis, acceptable? Likewise for the apparent entailment from (66b) to (78a), since (78b) is also acceptable.

(78)   a.   I would like to write my dissertation about a (certain) number, namely twelve.

   b.   I would like to write my dissertation about the number twelve.

In short, it is far from clear whether all of Moltmann's data actually support (MT7), or whether Moltmann's proposed explanation for apparent entailments like (63a,b) is consistent with the motivation for that thesis.

   A final question involves paraphrases of arithmetic equations.

(40)   Three and two is five.

(69a)   Three and two make five.

One thing worth noting about (40) and (69a) is that there appear to be similar kinds of examples not involving numerals. Consider (79a-c).[37]

(79)   a.   Ketchup and horseradish {is/makes} cocktail sauce.

   b.   Red and blue {is/makes} purple.

   c.   Twelve inches and two feet {is/makes} one yard.

---

[37] I omit discussion of the plural since there appears to be little agreement regarding their acceptability, and also because as Rothstein (2017) notes, there is cross-linguistic variation regarding whether number agreement is required in examples like (40).

As noted, these are also acceptable with 'make.' And, as with numerals, the terms flanking 'make' cannot be swapped.

(80)   a. ?? Cocktail sauce makes ketchup and horseradish.

     b. ?? Purple makes red and blue.

     c. ?? One yard makes twelve inches and two feet.

More generally, bare nouns like 'cocktail sauce,' color words like 'purple,' and measure phrases like 'one yard' appear to pattern just number expressions with respect to Moltmann's contrasts. This casts considerable doubt on the empirical legitimacy of "the *is* of calculation," in particular the claim that it is syncategorematic.

Indeed, (79) and (80) strongly suggest that a uniform, ideally *compositional*, semantic explanation is required, one which can account for all of these together. Furthermore, as we'll see in Section 3, the expressions in (79) and (80) are standardly assumed within linguistic semantics to have genuinely *referential* uses. Moreover, the same semantic mechanism said to be responsible for rendering them referential – *nominalization* – is also said to be responsible for rendering number expressions referential. All of this suggests that quasi-referentiality is not the only, let alone best, potential explanation of Moltmann's data.

3 Polymorphism and the Ontology of Number

The last section was framed around the Easy Argument, based on (11a,b).

(11a)   Jupiter has four moons.

(11b)   The number of Jupiter's moons is four.

This section is framed around a different puzzle, due to Hofweber (2005, 2016), also based on (11a,b). Whereas 'four' in (11a) appears to be an adjective, functioning as a modifier, 'four' in (11b) appears to be a name, functioning as a singular term. Yet if names characteristically refer and adjectives are characteristically non-referential, then how can a single expression—'four'—do *both*? Hofweber calls this FREGE'S OTHER PUZZLE (FOP).

FOP raises the question: How can a single expression serve multiple, possibly opposing semantic functions? Adding to this, note that number words have *many* uses beyond those witnessed in (11a,b), including:

(81)   a. Jupiter's moons are <u>four</u> (in number).   (predicative)

     b. Those are Jupiter's <u>four</u> moons.   (attributive)

     c. Jupiter has <u>four</u> moons.   (quantificational)

    d. The number of Jupiter's moons is <u>four</u>.   (specificational)

    e. Jupiter's moons number <u>four</u> (in total).   (verbal complement)

    f. <u>Four</u> is an even number.   (numeral)

    g. The number <u>four</u> is even.   (predicative numeral)

    h. Mary drank <u>four</u> ounces of water.   (measurement)

    i. Mary is contestant number <u>four</u>.   (ordinal)

Following Geurts (2006), the labels here—"predicative," "attributive," etc.—are intended to be useful mnemonics indicating how 'four' is used in the accompanying example. For instance, it plausibly functions as a predicate in (81a), similar to 'large' in (82).

(82)   Jupiter's moons are large (in size).

In contrast, it functions as a modifier in (81b), similar to 'large' in (83).

(83)   Those are Jupiter's large moons.

This is semantically significant because predicates and modifiers are generally assumed to have different semantic types, and thus *meanings*. Something similar can be said for other uses of 'four' in (81). Yet these different meanings are clearly not unrelated. After all, (81a-i) all intuitively convey something broadly number-related: how many moons Jupiter has, what a certain number is like, the amount of water Mary drank, or Mary's ordinal position among the contestants. Put differently, each involves a potential *application* of numbers: counting, calculating, measuring, and ordering.

    The primary contention of this section is that this important observation not only affords a natural resolution of FOP, but also a strengthened version of Hale's Argument. The empirical challenges facing substantivalism and adjectivalism highlight the need for an empirically adequate semantics for number words. What we want, ideally, is a *comprehensive* semantics, i.e. one which provides empirically plausible meanings for *all* occurrences of 'four' in (81). Yet a comprehensive semantics for number words is empirically adequate, I submit, only if it meets two desiderata: i) it compositionally provides meanings appropriate for all uses of number words, and ii) it explains how the meanings of those uses are systematically related. I will argue that the only kind of semantic theory capable of meeting both desiderata is one which not only recognizes that different occurrences of number words can serve different semantic functions, but also that number words sometimes function as singular terms, referring to numbers. Ultimately, then, the semantic evidence supports referentialism, and thus realism.

The rest of the section is laid out as follows. §3.1 introduces FOP, discusses four potential solutions to it, and proposes a resolution naturally suggested by widespread assumptions within contemporary semantic theory. Along the way, I introduce certain technical notions crucial to meeting the two desiderata mentioned, namely polymorphism and type-shifting. In §3.2, I divide extant polymorphic analyses into two kinds, showing how both provide meanings appropriate for (81a-i), while also explaining how those meanings are related. Specifically, on both analyses, there is an element common to each of the meanings witnessed in (81a-i), namely a number. Ultimately, this provides a strengthened form of Hale's Argument, elaborated in §3.3. I conclude the monograph in §3.4, where I summarize the resulting dialectic and its implications for ontology.

3.1 Frege's Other Puzzle

As mentioned, FOP is framed around Frege's (11a,b), where 'four' appears to serve different, and indeed opposing, semantic functions. As Hofweber (2016, p. 115–116) explains, this is puzzling:

> On the one hand, "four" occurs as an adjective [in (11a)], which is to say that it occurs grammatically in sentences in a position that is commonly occupied by adjectives ... similar to "green" in [(84)].
>
> (84)   Jupiter has green moons.
>
> On the other hand, "four" occurs as a singular term [in (11b)], which is to say that it occurs in a position that is commonly occupied by paradigmatic cases of singular terms, ... [so that] "four" [in (11b)] seems to be just like "Wagner" in
>
> (15)   The composer of *Tannhäuser* is Wagner.
>
> and both of these statements seem to be identity statements, ones with which we claim that what four singular terms stand for is identical.
>
> But that number words can occur both as singular terms and as adjectives is puzzling. Usually adjectives cannot occur in a position occupied by a singular term, and the other way round, without resulting in ungrammaticality and nonsense. To give just one example, it would be ungrammatical to replace "four" with "the number of moons of Jupiter" in [(85)]:
>
> (85)   Jupiter has the number of Jupiter's moons moons.
>
> This ungrammaticality results even though "four" and "the number of moons of Jupiter" are both apparently singular terms standing for the same object in [(85)]. So, how can it be that number words can occur both as singular terms and as adjectives, while other adjectives cannot occur as singular terms, and other singular terms cannot occur as adjectives?
>
> Even though Frege raised this question more than 100 years ago, I dare say that no satisfactory answer has ever been given to it.

Hofweber's comments suggest characterizing FOP as an inconsistent tetrad.

Same Expression: Both occurrences of 'four' in (11a,b) are witness to the same expression.

Anti-Substantivalism: 'four' in (11a) is an adjective, acceptably functioning as a modifier.

Anti-Adjectivalism: 'four' in (11b) is a numeral, acceptably functioning as a singular term.

Single Function: Different occurrences of the same expression cannot acceptably serve different semantic functions. In particular, different occurrences of adjectives or names cannot acceptably function as both singular terms and modifiers.

Although each of these claims seems plausible, they are jointly inconsistent. Thus, at least one must be rejected. But which?

3.1.1 Four Potential Solutions

Since FOP is an inconsistent tetrad, four potential solutions arise, depending on which thesis one rejects. Obviously, substantivalism denies Anti-Substantivalism, whereas adjectivalism denies Anti-Adjectivalism. However, we have seen that both kinds of strategies face empirical difficulties. For example, if substantivalism were correct, then we would expect singular terms coreferential with the numeral 'four,' such as 'the number of Jupiter's moons' on Frege's analysis, to be acceptably substitutable for that numeral. Yet as Hofweber's (85) suggests, this is not the case. Conversely, if adjectivalism were correct, then we would not expect 'four' in (11b) to permit existential generalization of the sort apparently characteristic of singular terms. For example, (86a) is clearly not entailed by (11a).

(86)   a. ?? Jupiter has something moons, namely four.

         b. The number of Jupiter's moons is something, namely four.

In contrast, it appears that (11b) entails (86b).

A different potential solution would be to deny Same Expression. On this strategy, the different occurrences of 'four' in (11a,b) are HOMONYMS, i.e. different expressions which happen to be spelled and pronounced alike. Generally, we do not expect homonyms, such as the noun 'fire' and the verb 'fire,' to be acceptably intersubstitutable. Consider (87b), paralleling (85).

(87)   a. The rapid oxidation of combustible materials is fire.

         b. Let's {fire/??the rapid oxidation of combustible materials} John.

Perhaps this could explain why (85) is similarly unacceptable. On the other hand, because homonyms are typically spelled and pronounced alike as a matter of *historical accident*, we do not expect their meanings to be related. Thus, the problem with this strategy is that the occurrences of 'four' in (11a,b) clearly are semantically related. Indeed, both tell us something about the cardinality of Jupiter's moons.

Thus, the solution defended here is to deny Single Function: different occurrences of the same expression can in fact serve different semantic functions. Specifically, adjectives and names can both acceptably function as singular terms and as modifiers. For example, consider (88), where 'green' functions referentially, unlike its occurrence in (84).

(88)   Green is a secondary color.

Within linguistic semantics, color words like 'green' are usually assumed to be adjectives,[38] which come to function referentially through NOMINALIZATION.[39] One way of nominalizing adjectives in English is to form gerunds and infinitives like those in (89a), due to Chierchia (1984). Another is to use nominalizing suffixes, e.g. '-ness' or '-ity,' as with (89b).

(89)   a.   {??Nice/Being nice/To be nice} is nice.

       b.   {??Stupid/Stupidness/Stupidity} is annoying.

A third way, evidently witnessed by color words, is to form a name directly from its lexical root.[40] As a singular term, 'green' in (88) thus plausibly represents a counterexample to Single Function.

Now consider the name 'Trump,' which functions referentially in (90a), but instead as a modifier in (90b), similar to 'wealthy.'

(90)   a.   Trump is a controversial president.

       b.   Some {Trump/wealthy} supporters wear MAGA hats.

Thus, (90b) plausibly represents another counterexample to Single Function.

Of course, adequately solving FOP requires more than observing that Single Function has counterexamples. For one thing, not all adjectives can be nominalized in a manner similar to 'green,' as evidenced by (89a,b). So why think that numerals are more like 'green' than 'nice' or 'stupid' in this respect? Furthermore, names used as modifiers typically have a relational character. For example, 'Trump supporter' in (90b) is roughly synonymous with 'supporter of Trump.' Yet 'four moons' in (11b) is not roughly synonymous with 'moons of four'. Why, then, think that numerals can similarly function as modifiers?

---

[38] See e.g. Kennedy and McNally (2010) and McNally and de Swart (2011).

[39] See e.g. Chierchia (1984), Partee (1986b), and McNally and de Swart (2011).

[40] See Alexiadou (2017).

Another concern with denying Single Function can be extrapolated from Hofweber (2007). Consider the following, intuitively attractive principle:

Commitment: Sentences which are made true in exactly the same circumstances carry the same ontological commitments.

The intuitive rationale for Commitment should be evident: sentences which are made true in exactly the same circumstances make the same ontological demands on the world, so to speak. That is, what is required from the world to make one true must also be what is required to make the other true. Now, consistent with §2.1, suppose that (11a) does not imply commitment to numbers, while (11b) does. Then it would appear that denying Single Function also requires denying Commitment, contrary to intuition, or possibly even empirical fact, assuming (11a,b) are equivalent.[41]

Thus, as a potential solution to FOP, denying Single Function leaves certain important questions open. These include:

Categorization: If the different occurrences of 'four' in (11a,b) are witness to the same expression, then it is presumably either a numeral or an adjective. If so, then which is it?

Different Functions: If 'four' functions non-referentially in (11a) but referentially in (11b), then how can the same expression serve different, and indeed opposing, semantic functions?

Polysemy: Since expressions serving different semantic functions have different meanings, how are (11a,b) semantically related, both telling us something about the cardinality of Jupiter's moons?

Ontology: If 'four' functions non-referentially in (11a) but referentially in (11b), then how can (11a,b) be equivalent if (11b) but not (11a) implies commitment to numbers?

To see what answers to these questions might look like, we will first need to look at expressions which, like color words, can perform different but related semantic functions in different syntactic environments. Within linguistic semantics, such expressions are known as POLYMORPHIC.

3.1.2 A Brief Overview of Polymorphism and Type-Shifting

Polymorphic expressions are those which can take on different semantic types, and thus meanings, in different syntactic environments. To illustrate, consider INTERSECTIVE ADJECTIVES like 'green.' These typically have predicative and attributive forms, respectively witnessed in (91a,b).

---

[41] Modulo the issue raised in §2.1.1.

(91)   a.   That new car is <u>green</u>.

        b.   That <u>green</u> car is new.

Intersective adjectives are so-called because they, unlike e.g. 'former,' give rise to the characteristic entailment in (92a).[42]

(92)   a.   That is a green car $\vDash$ That is a car and it is green

        b.   That is a former senator $\nvDash$ That is a senator and she is former

In other words, when functioning as a modifier, 'green' intuitively denotes the intersection of two sets: the set denoted by the noun modified ('car'), and the set denoted by the corresponding predicate ('is green').

The important observation is that because 'green' functions as a predicate in (91a) and as a modifier in (91b), it also has different meanings in those examples. Nevertheless, those meanings are clearly related: whereas the former is true of green things, the latter restricts the extension of a noun to things which are green. Standardly, this correspondence in meaning is captured through TYPE-SHIFTING. Specifically, it is assumed that 'green,' like all polymorphic expressions, has a certain lexical meaning, in this case witnessed in predicative uses, and given in (93a). A meaning appropriate for attributive uses is then derivable from the latter via the type-shifting principle in (93a), which Landman (2004)'s calls 'ADJUNCT.'[43]

(93)   a. $\llbracket\text{green}\rrbracket = \lambda x.\ \text{green}(x)$

        b. $\text{ADJUNCT} = \lambda P.\lambda Q.\lambda x.\ P(x) \wedge Q(x)$

        c. $\text{ADJUNCT}(\lambda x.\ \text{green}(x)) = \lambda Q.\lambda x.\ \text{green}(x) \wedge Q(x)$

        d. $\llbracket\text{green car}\rrbracket = \lambda x.\ \text{green}(x) \wedge \text{car}(x)$

According to (93a), the lexical meaning of 'green' is that of a predicate true of all and only green things. This can then combine with ADJUNCT to return an intersective modifier, as suggested in (93c). The prediction is that 'green car' denotes those things which are both green and a car, cf. (93d), thus capturing the entailment in (92a).

As a second illustration, consider again nominal uses of 'green' like (88).

(88)   Green is a secondary color.

As mentioned, such uses are standardly assumed to arise through nominalization. Semantically speaking, nominalization involves naming the property expressed by a given predicate. In the case of 'green,' for instance, it involves naming the property of being green. On standard accounts,[44] this is captured

---

[42] See e.g. Partee (2004) and Kennedy (2012).

[43] See Partee (2004).

[44] See e.g. Chierchia (1984) and McNally and de Swart (2011).

through another type-shifting operation, namely Partee (1986b)'s 'NOM,' defined in (94a), where '$^{\cap}$' is Chierchia (1984)'s nominalization-operation.

(94)   a. NOM = $\lambda P.\ ^{\cap}[\lambda x.\ P(x)]$

  b. NOM($\lambda x.\ \text{green}(x)$) = $^{\cap}[\lambda x.\ \text{green}(x)]$

Properties play two roles in Chierchia (1984)'s PROPERTY THEORY, corresponding roughly to Frege (1951)'s distinction between concept and object. First, they may be predicated of an object, as witnessed in predicative uses. Secondly, they may be viewed as entities in their own right, or what Chierchia calls INDIVIDUAL PROPERTY CORRELATES. The $^{\cap}$-operation takes a predicate as argument and returns a singular term whose referent is the corresponding individual property correlate. Thus, as before, predicative and nominal uses of 'green' are systematically related: whereas the former expresses a property, the latter names that property, viewed as an entity.

More generally, type-shifting principles shift the semantic type, and thus meaning, of an expression. For instance, whereas ADJUNCT shifts the meaning of 'green' from that of a predicate (type $\langle e, t \rangle$) to that of an intersective modifier (type $\langle \langle e, t \rangle, \langle e, t \rangle \rangle$), NOM instead shifts the predicate into a singular term (type e). What's more, these same type-shifting principles are potentially applicable to *any* expression of the appropriate type. For example, ADJUNCT would apply to the predicative form of any intersective adjective to generate a meaning appropriate for the corresponding attributive form. In this way, ADJUNCT codifies an important semantic generalization, namely that intersective adjectives are generally POLYSEMOUS, having different but related predicative and attributive meanings.

Likewise for polymorphic expressions more generally. Since they take on different meanings via independently motivated and well-attested type-shifting principles, those meanings are systematically related. Consequently, polymorphic expressions are generally polysemous.

3.1.3 Rejecting Single Function

Number words resemble prototypically polymorphic expressions, such as color words and measure phrases, in important respects. For example, all three have predicative, attributive, specificational, and referential uses:

(95)   a. Jupiter's moons are four (in number).

  b. Jupiter's moons are green (in hue).

  c. Mary's hammers are four pounds (in weight).

(96)   a. Those are (Jupiter's) four moons.

  b. Those are (Jupiter's) green moons.

    c. Those are (Mary's) four-pound hammers.

(97) a. The number of Jupiter's moons is four.

    b. The color of Jupiter's moons is green.

    c. The weight of Mary's hammers is four pounds.

(98) a. Four is a number.

    b. Green is a color.

    c. Four pounds is a weight.

Given these similarities, it is straightforward to formulate analogs of FOP for color words and measure phrases. For example, assuming that various occurrences of 'green' and 'four pounds' in (95)–(98) are witness to the same expression, we may reasonably wonder how they are apparently capable of functioning referentially, predicatively, and attributively. Yet the analogs of substantivalism and adjectivalism are hardly attractive in these cases. There is little intuitive plausibility to the claim that 'green' serves the *same* referential or non-referential function in (95b)–(98b), for example. So why should things be any different for 'four'?

    Indeed, it is highly plausible that all three kinds of expressions are subject to the same type-shifting principles,[45] thus explaining why they pattern alike with respect to the uses noted. If so, then different occurrences of the same expression—'four,' 'green,' or 'four pounds'—can in fact serve different semantic functions. In other words, semantic orthodoxy recommends rejecting the analog of Single Function for all three kinds of expressions. Seen this way, FOP is only puzzling if we neglect the polymorphic nature of number words,[46] and indeed a variety of other kinds of expressions in English.

    How might this observation offer answers to our four questions above? Consider first Categorization: Is 'four' a numeral or an adjective? As we will see in §3.2, there are two kinds of standard polymorphic analyses available, which differ in the lexical meaning assumed for 'four': it is either that of a numeral referring to a number, or else that of an adjective true of pluralities of four things. Thus, the answer to Categorization will depend on which of these analyses one adopts. In either case, however, it is possible to derive meanings appropriate for all uses of 'four,' including those in (11a,b), using roughly the same type-shifting principles. Consequently, independent of which of these two kinds of analyses is correct, Single Function is false.

---

[45] See e.g. Partee (1992) for adjectives, and Rothstein (2017) for measure phrases.

[46] Contra Hofweber (2016), who argues that number words cannot be polymorphic. For a response, see Snyder et al. (ms.).

Next, consider again Different Functions: How can 'four' function both referentially and non-referentially? Like all polymorphic expressions, number words take on different semantic types, and thus functions, via type-shifting. The key difference between the two kinds of analyses mentioned is that on one, certain referential uses are witness to the lexical meaning of 'four,' while on the other kind, certain non-referential uses are witness to its lexical meaning. But on both kinds of analyses, 'four' takes on the *same range* of meanings, including those appropriate for (11a,b). In this respect, number words resemble color words and measure phrases, for instance.

Now consider Polysemy: How are the different meanings witnessed in (11a,b) related? The characteristic feature of polymorphic expressions is that they can take on different meanings which are either witnessed to, or else derivable from, a single lexical meaning, thanks to type-shifting, thus explaining their polysemy. Likewise for number words. Indeed, we'll see in §3.2 that meanings appropriate for 'four' in (11a,b) are obtainable on both kinds of polymorphic analyses, and that both meanings have an element in common which plausibly explains why (11a,b) both intuitively convey something about how many moons belong to Jupiter.

Finally, consider Ontology: How can (11a,b) be equivalent if (11b) implies commitment to numbers while (11a) does not? The answer, in brief, is that (11a,b) *both* carry ontological commitment to numbers, despite 'four' in those examples serving different semantic functions. In fact, on both kinds of polymorphic analyses mentioned, if numbers did not exist, then neither (11a) nor (11b) could be true. Likewise for *all* uses of number words noted in (81a-i), it turns out: without numbers, none of them could be true. Ultimately, this highlights a semantically significant fact about the meanings number words can take on: they are related in virtue of sharing of a certain element in common, namely a number. This insight will play a crucial role in §3.4, where I consider nominalist responses to Hale's Argument.

3.2 Two Kinds of Polymorphic Analyses

As mentioned, the widespread assumption within linguistic semantics is that number words, such as 'four,' are polymorphic. The purpose of this section is to group extant analyses into two kinds, and to show how both are capable of generating meanings appropriate for the various uses of 'four' noted in (81),[47] in keeping with the solution to FOP just proposed.

---

[47] I am ignoring additional uses of 'four' like those in (ia,b), for space. For analyses consistent with the semantics developed here, see Rothstein (2013, 2017) for complex cardinals, and Snyder and Barlew (2019) and Snyder (2020) for fractions.

On the first kind of analysis, the lexical meaning of 'four' is that of a numeral, and meanings appropriate for all other uses are derivable from it via type-shifting. On the second, the lexical meaning of 'four' is that of an adjective or determiner, and all other meanings are derivable from it via type-shifting. Due to the resemblance to Dummett (1991)'s two strategies, I call the first kind of analysis POLYMORPHIC SUBSTANTIVALISM (PS), and the second kind POLYMORPHIC ADJECTIVALISM (PA). Versions of PS have been defended by Landman (2003, 2004), Scontras (2014), and Snyder (2017, 2020), while versions of PA have been suggested or defended by Partee (1986b), Geurts (2006), Rothstein (2013, 2017), and Kennedy (2015).

To keep things manageable, I will recruit one analysis to represent PS, and one to represent PA. Specifically, Landman (2003, 2004)'s analysis will represent PS, while Rothstein (2013, 2017)'s analysis will represent PA.[48] According to the former, the lexical meaning of 'four' is given in (99a), while according to the latter, it is instead given in (99b).

(99)   a. $[\![\text{four}]\!] = 4$ (PS)

       b. $[\![\text{four}]\!] = \lambda x.\ \mu_\#(x) = 4$ (PA)

Here, '4' represents the number four, and '$\mu_\#$' is a MEASURE FUNCTION, i.e. a function from entities to numbers. Specifically, '$\mu_\#$' measures cardinality, mapping pluralities to numbers representing how many countable individuals are part of that plurality. Thus, according to PS, the lexical meaning of 'four' is that of a numeral referring to the number four, while according to PA, it is instead that of a predicative adjective true of pluralities having exactly four countable parts.

As we will see, it is possible to derive meanings appropriate for (81a-i) from both (99a,b), using roughly the same stock of type-shifting principles. Thus, PS and PA may be fruitfully seen as different views about the basic semantic function of number words: lexically, they are either referential expressions (PS), or else non-referential expressions (PA). In either case, however, they take on numerous other meanings, thanks to type-shifting, and all of these, it will turn out, share an element in common: a number.

(i)   a. Four hundred troops surrounded the fort.   (complex cardinal)

       b. Mary ate four fifths of the pie.   (fractional)

---

[48] My choice of representatives is motivated largely by two considerations. First, unlike some of the other analyses cited, Landman's and Rothstein's analyses are explicitly formulated to cover a wide range of potential meanings. Secondly, the derivations proposed are very similar, thus simplifying the presentation here considerably. The chief differences between these and the other polymorphic analyses mentioned are negligible, in that the arguments presented in §3.3 would go through regardless of this choice.

To be clear, my purpose here is not to *adjudicate* between these two kinds of analyses. Rather, my aim is to show that independent of whether one views 'four' as lexically referential or lexically non-referential, numbers will be needed to explain how the various meanings number words can take on are related. Thus, independent of which kind of polymorphic analysis is correct, the best available semantic evidence supports realism.

To facilitate discussion, I will group the uses of 'four' in (81) into four kinds: CARDINAL USES like (81a-e), which convey cardinal information; ARITHMETIC USES like (81f,g), which convey arithmetic information; MEASUREMENT USES like (81g), which convey information about measurement; and ORDINAL USES like (81h), which convey ordinal information. For reasons which will become evident, I begin with cardinal uses.

3.2.1 Cardinal Uses

Intuitively, (81a-e) all report something about the cardinality of Jupiter's moons, i.e. they tell us how many moons it has.

(81a) Jupiter's moons are <u>four</u> (in number).   (predicative)

(81b) Those are Jupiter's <u>four</u> moons.   (attributive)

(81c) Jupiter has <u>four</u> moons.   (quantificational)

(81d) The number of Jupiter's moons is <u>four</u>.   (specificational)

(81e) Jupiter's moons number <u>four</u> (in total).   (verbal complement)

To make sense of this, we will first need a background semantics for count nouns, such as 'moon.' Here, I follow Link (1983)'s mereological analysis, according to which the singular count noun 'moon' denotes a set of individuated, or ATOMIC, moons. Technically, an atomic F is anything such that it but none of its proper parts satisfies F.[49] Thus, an atomic moon is anything which, vagueness aside, is itself a moon but which has no proper parts which are moons. In a sense, then, atoms serve as *units* of a given property: they are the "smallest" things having that property.

Plural nouns like 'moons' then denote the closure of atoms under sum-formation, represented in (100b) by '*.'

(100) a. $[\![\text{moon}]\!] = \lambda x. \text{moon}(x)$

       b. $[\![\text{moons}]\!] = \lambda x. {}^*\text{moon}(x)$

       c. $[\![\text{the}]\!] = \lambda P. \sigma x[P(x)]$

       d. $[\![\text{the moon}]\!] = \sigma x[\text{moon}(x)]$

       e. $[\![\text{the moons}]\!] = \sigma x[{}^*\text{moon}(x)]$

---

[49] Cf. Krifka (1989) and Link (1998).

Thus, 'moons' denotes all atomic moons plus all mereological sums ($\sqcup$) of those atoms. Following Sharvy (1980), 'the' is then analyzed as a maximality-operation, codified in (100c) as 'σ.' This binds a variable to form a singular term referring to the maximal salient thing satisfying the accompanying description. When combining with a singular count noun like 'moon,' 'σ' will pick out a unique salient moon (if there is one), cf. (100d). When combining with plural nouns like 'moons,' on the other hand, 'σ' will instead pick out the maximal salient sum of moons (if there is one), cf. (100e).

Suppose we have three atomic moons: a, b, and c. Ordering these by parthood ($\sqsubseteq$) results is an ATOMIC JOIN SEMILATTICE structure like the following, where lines represent parthood, defined as $a \sqsubseteq b \leftrightarrow a \sqcup b = b$, and dots represent denotations.

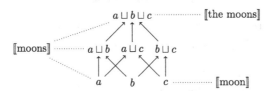

The bottom elements of the structure are atoms, while the remaining elements are sums. Within Link's semantics, these enjoy the same semantic type and ontological status. They are individuals, though different *sorts* of individuals: ATOMIC INDIVIDUALS and PLURAL INDIVIDUALS, respectively.

On both PS and PA, predicative uses of 'four' like (81a) denote sums having a certain number of atomic parts ($\sqsubseteq_{At}$), defined as $a \sqsubseteq_{At} b \leftrightarrow a \sqsubseteq b \wedge \text{ATOM}(a)$, where 'ATOM' is a metalinguistic predicate true of atoms.[50] Specifically, 'four' in (81a) has the meaning in (101), where '$\mu_\#$' is a function mapping individuals to numbers representing their atomic parts.

(101) $[\![\text{four (in number)}]\!] = \lambda x.\ \mu_\#(x) = 4$

More generally, the resulting picture looks as follows:

$[\![\text{three (in number)}]\!]$ .... $a \sqcup b \sqcup c$

$[\![\text{two (in number)}]\!]$ .... $a \sqcup b \quad a \sqcup c \quad b \sqcup c$

$[\![\text{one (in number)}]\!]$ .... $a \qquad b \qquad c$

---

[50] Realistically, atomicity should be relativized to a property and a sort (Krifka (1989)), and probably also a context (Rothstein (2017),Snyder and Shapiro (2020)).

So, 'one (in number)' is true of all F-atoms, 'two (in number)' is true of all F-sums having exactly two atomic parts, etc.

The crucial difference between PS and PA lies in how these predicative meanings arise. On PA, the lexical meaning of 'four' is precisely the cardinality predicate in (101). In contrast, according to PS, this predicative meaning arises via type-shifting. Specifically, it is generated through a type-shifting operation I dub 'NUM,' defined in (102b).

(102)　a.　$[\![\text{four}]\!] = 4$

　　　b.　$\text{NUM} = \lambda n.\lambda x.\ \mu_{\#}(x) = n$

　　　c.　$\text{NUM}(4) = \lambda x.\ \mu_{\#}(x) = 4$

Thus, according to PS, combining the lexical meaning of 'four' in (102a) with 'NUM' returns the same cardinality predicate given in (101), cf. (102c). It is in this respect that PS and PA differ with respect to the basic semantic function of 'four': according to the former, it is lexically referential, while according to the latter, it is lexically non-referential.

With cardinality predicates in place, we may now compositionally derive intuitively correct truth-conditions for (81a). Specifically, following Barker (1998), assume the possessive morpheme has the meaning in (103a).

(103)　a.　$[\![\text{-'s}]\!] = \lambda x.\lambda P.\lambda y.\ P(y) \wedge R(y,x)$

　　　b.　$[\![\text{Jupiter's}]\!] = \lambda P.\lambda y.\ P(y) \wedge R(y,j)$

　　　c.　$[\![\text{Jupiter's moons}]\!] = \lambda y.\ {}^{*}\text{moon}(y) \wedge R(y,j)$

Here, free 'R' indicates that its interpretation is provided by context. Accordingly, 'Jupiter's moons' is a predicate true of pluralities of moons bearing some contextually-supplied relation (e.g. belonging) to Jupiter. Notice, however, this has the wrong semantic type to combine with predicative 'four,' given (101)/(102c). This mismatch is readily fixed, however, thanks to Partee (1986b)'s IOTA, defined in (104a).

(104)　a.　$\text{IOTA} = \lambda P.\ \sigma x[P(x)]$

　　　b.　$\text{IOTA}(\lambda y.\ {}^{*}\text{moon}(y) \wedge R(y,j)) = \sigma y[{}^{*}\text{moon}(y) \wedge R(y,j)]$

Thus, IOTA codifies the (Fregean) meaning of 'the' from above: it takes a predicate and returns a singular term referring to the maximal salient entity satisfying that predicate. For example, it combines with 'Jupiter's moons' in (103c) to return a singular term referring to the maximal plurality of moons belonging to Jupiter. This can then serve as argument to the predicate in (101)/(102c). Hence, (81a) will be true just in case that plurality has exactly four atomic parts, i.e. just in case Jupiter has exactly four moons.

With predicative meanings in place, meanings appropriate for other cardinal uses of 'four' can be generated via additional type-shifting operations. For example, consider (105a,b).

(105)   a. {All/No} four moons are gathered close to Jupiter.

       b. {All/No} green leaves are scattered on the floor.

For compositional reasons, 'four' in (105a) needs to function as a modifier, similar to 'green' in (105b). The crucial observation is that, as a modifier, 'four' is *intersective*,[51] as witnessed by the fact that it, like 'green,' licenses the characteristic entailment in (106).

(106)   Those are Jupiter's four moons $\models$ Those are Jupiter's moons and they are four (in number)

Thus, cardinal modifiers are intersective, denoting the intersection of the things in the denotation of the modified noun and the things having the cardinality property in question.

In order to capture this fact, the attributive meaning witnessed in (105a) is derived from predicative 'four' via ADJUNCT, repeated in (107a).

(107)   a. $\text{ADJUNCT} = \lambda P.\lambda Q.\lambda x.\, P(x) \wedge Q(x)$

       b. $\text{ADJUNCT}(\lambda x.\, \mu_{\#}(x) = 4) = \lambda Q.\lambda x.\, \mu_{\#}(x) = 4 \wedge Q(x)$

       c. $[\![\text{four moons}]\!] = \lambda x.\, \mu_{\#}(x) = 4 \wedge {}^{*}\text{moon}(x)$

Thus, applying ADJUNCT to predicative 'four' returns an intersective modifier which, when combined with 'moons,' delivers a predicate true of pluralities of moons having exactly four atomic parts, cf. (107c).

The intuitively correct prediction is that (81b) is true just in case 'those' deictically references such a plurality (belonging to Jupiter). Notice also that entailment in (106) falls out immediately. Furthermore, assuming 'all' and 'no' have the GQT-denotations given in (108a,b) (see §2.3.1),

(108)   a. $[\![\text{all}]\!] = \lambda P.\lambda Q.\, P \subseteq Q$

       b. $[\![\text{no}]\!] = \lambda P.\lambda Q.\, P \cap Q = \varnothing$

(105a) will thus be true with 'all' if the set of pluralities consisting of exactly four moons are a subset of the things which are gathered close to Jupiter, while it will be true with 'no' if the two sets are disjoint.

Finally, we turn to quantificational uses of 'four' like (81c). For guidance, it will be useful to compare the latter to Hofweber's (84).

(84)   Jupiter has green moons.

---

[51] Cf. Partee (1986b), Landman (2003, 2004), Rothstein (2017), and Snyder (2017).

Clearly, (84) would not be false if Jupiter also happened to have red moons, for instance. Rather, all that's apparently required is that Jupiter has *some* green moons. This is accounted for by adopting another of Partee (1986b)'s type-shifters, namely A, defined in (109a).

(109)   a. $A = \lambda P.\lambda Q.\exists x.\ P(x) \wedge Q(x)$

       b. $[\![\text{green moons}]\!] = \lambda x.\ \texttt{green}(x) \wedge {}^*\texttt{moon}(x)$

       c. $A(\lambda x.\ \texttt{green}(x) \wedge {}^*\texttt{moon}(x)) = \lambda Q.\exists x.\ \texttt{green}(x) \wedge {}^*\texttt{moon}(x) \wedge$
          $Q(x)$

Thus, assuming that 'green' in 'green moons' is an intersective modifier whose meaning arises in the manner suggested in §3.1.3, thus resulting in (109b), combining this predicate with A will then result in (109c). Accordingly, (84) will be true if at least one plurality of green moons belongs to Jupiter, i.e. if Jupiter has *some* green moons.

PS and PA both generate a meaning appropriate for 'four' in (81c) similarly. Specifically, applying A to 'four moons' in (107c) returns (110).

(110) $A(\lambda x.\ \mu_{\#}(x) = 4 \wedge {}^*\texttt{moon}(x)) = \lambda Q.\exists x.\mu_{\#}(x) = 4 \wedge {}^*\texttt{moon}(x) \wedge Q(x)$

Accordingly, (81c) will be true if there is at least one plurality of exactly four moons belonging to Jupiter. Notice that this would be true even if Jupiter had more than four moons, as there would still be at least one plurality of four moons in that case.[52]

In sum, both analyses are capable of generating meanings seemingly appropriate for a variety of cardinal uses, and they do so through similar semantic means (we will return to (81d,e) in §3.2.3). The only substantial difference lies in their explanation of predicative cardinal meanings. Whereas these are lexical on PA, they are instead derived on PS. The situation is exactly reversed for arithmetic uses, as we'll see presently.

3.2.2 Arithmetic Uses

We now turn to arithmetic uses like (81f,g).

(81f)   <u>Four</u> is an even number.   (numeral)

(81g)   The number <u>four</u> is even.   (predicative numeral)

As with cardinal uses, arithmetic uses reveal the basic ideological difference between PS and PA. Specifically, whereas meanings appropriate for (81f) are lexical according to PS, they are derived via type-shifting on PA.

---

[52]   That is, (110) predicts *lower-bounded* ("at least *n*") truth-conditions, where two-sided ("exactly *n*") interpretations are said to arise via a Gricean scalar implicature (see Horn (1972)). This is controversial, however (see Geurts (2006) and Kennedy (2015)).

Let's begin with PS. According to it, recall, the lexical meaning of 'four' is that of a numeral naming the number four.

(99a) $[\![four]\!] = 4$

This is the meaning of the numeral 'four' in (81f). In terms of Link (1983)'s ontology, numbers are *atomic individuals*, just like Mary, this table, or New York. Unlike the latter, however, numbers form a designated semantic type, representing a distinct ontological category. Specifically, numbers are type n, or a sub-type of entities (type e). Thus, following Landman (2003, 2004), we may employ sorted variables 'n,' 'n',' etc. to range specifically over entities of that type. They are the denotation of monadic, arithmetic 'number' (see §2.2.2) witnessed in (81f), the meaning of which is given in (111).

(111) $[\![number_a]\!] = \lambda n.\, \texttt{number}(n)$

Hence, (81f) will be true if the referent of 'four' is in this set and is even.

As atomic individuals, we can count numbers, quantify over them, and deictically and anaphorically refer to them.

(112)   a.   There are three numbers inclusively between five and seven: five, six, and seven.

   b.   Some even number is prime, namely two.

   c.   [Pointing at the blackboard:] Mary divided that number by ten and by two. Both of those numbers are even.

Numbers can also stand in various arithmetic relations to each other, such as being less than or being the arithmetic sum of. Following Landman (2004), these are definable as follows.

(113)   a. $[\![less\ than]\!] = \lambda n.\lambda n'.\, n' < n$

   b. $[\![plus]\!] = \lambda n.\lambda n'.\, n' + n$

   c. $[\![equals]\!] = \lambda n.\lambda n'.\, n' = n$

Thus, (114) is true if adding four to three results in the number seven.

(114)   Four plus three equals seven.

More generally, our arithmetic talk is not especially different from our non-arithmetic talk, semantically speaking. Rather, as its surface syntax suggests, it is about arithmetic objects and their arithmetic properties.

Likewise on PA, only the referents of numerals are generated via type-shifting from lexical predicates like (99b).

(99b) $[\![four]\!] = \lambda x.\, \mu_{\#}(x) = 4$

Specifically, meanings for numerals are generated via nominalization, similar to color words (see §3.1.2). Thus, just as a meaning appropriate for nominal uses of 'green' is generated via Partee (1986b)'s NOM, returning the individual property correlate of being green, similarly a meaning appropriate for the numeral 'four' results from applying NOM to (99b), returning the individual property correlate of being four in number.

(115) $NOM(\lambda x.\ \mu_\#(x) = 4) = {}^\cap[\lambda x.\ \mu_\#(x) = 4]$

On the other hand, we saw that (81a) is true only if 'four' references a *number*, and (115) will not alone guarantee that certain individual property correlates are numbers. Rather, this must be stipulated. In particular, Rothstein (2013, 2017) posits the schematic equation in (116).

(116) $n = {}^\cap[\lambda x.\ \mu_\#(x) = n]$

As before, 'n' here ranges over numbers. The effect of (116) is thus to *identify* numbers with certain individual property correlates.

With numbers in place, a predicative meaning appropriate for 'four' in (81g) can then be derived via type-shifting on both analyses.[53] First, notice that when forming definites, 'the number ...' can occur with certain relational expressions like 'less than' and 'identical to.'

(117)   the number {less than/which is/identical to} four

Intuitively, 'the number four' is coreferential with 'the number which is identical to four,' i.e. both function as referential expressions picking out a number coreferential with the numeral 'four.' This can be captured by another of Partee (1986b)'s type-shifting operations, namely, IDENT, defined in (118a), which takes a singular term and returns an identity predicate true of all things identical to its referent.

(118)   a. $IDENT = \lambda x.\lambda y.\ y = x$

   b. $IDENT(4) = \lambda n.\ n = 4$

   c. $ADJUNCT(\lambda n.\ number(n)) = \lambda Q.\lambda n.\ number(n) \wedge Q(n)$

   d. $[\![number_a\ four]\!] = \lambda n.\ number(n) \wedge n = 4$

   e. $[\![the\ number_a\ four]\!] = \sigma n[number(n) \wedge n = 4]$

Thus, applying IDENT to the numeral 'four' returns an identity predicate true of all and only numbers identical to four, cf. (118b). Next, we apply ADJUNCT to arithmetic 'number' in (111), thus returning the intersective modifier in (118c).

---

[53] Cf. Snyder (2017).

Combining this with the identity predicate in (118b) thus results in a predicate true of numbers identical to four, cf. (118d). Applying 'the' from above, 'the number four' will thus be a singular term referring to the unique number which is identical to four, i.e. four itself, cf. (118e).

The welcome result is that the numeral 'four' and 'the number four' are coreferential singular terms. Consequently, (81f,g) are equivalent, both being true if their referent has the arithmetic property of being even.

3.2.3 Measurement Uses

With numbers in place, we can now provide meanings appropriate for 'four' in measure phrases such as 'four ounces of water.'

(81h)   Mary drank <u>four</u> ounces of water.   (measurement)

On both PS and PA, measure phrases are complex, consisting of a numeral such as 'four,' a MEASURE NOUN such as 'ounce' or 'pound,' and a SUBSTANCE NOUN such as 'water' or 'boxes.' As the label indicates, substance nouns denote substances, i.e. mass quantities or pluralities.[54] Intuitively, measure phrases measure substances according to a dimension of measurement provided by the measure noun. For example, whereas (119a) measures water by volume, (119b) measures boxes by weight.

(119)   a.   That is four ounces of water.

        b.   Those are four pounds of boxes.

This is captured by the following analysis, due to Rothstein (2016).

(120)   a. $[\![\text{ounce}]\!] = \lambda n.\lambda x.\ \mu_{oz,vol}(x) = n$

        b. $[\![\text{four ounces}]\!] = \lambda x.\ \mu_{oz,vol}(x) = 4$

        c. $\text{ADJUNCT}(\lambda x.\ \mu_{oz,vol}(x) = 4) = \lambda Q.\lambda x.\ \mu_{oz,vol}(x) = 4 \wedge Q(x)$

        d. $[\![\text{of water}]\!] = \lambda x.\ \text{water}(x)$

        e. $[\![\text{four ounces of water}]\!] = \lambda x.\ \mu_{oz,vol}(x) = 4 \wedge \text{water}(x)$

As before, '$\mu_{oz,vol}$' is a measure function. In this case, however, it is relativized to two parameters: '$_{oz}$,' indicating a unit of measurement (an ounce), and '$_{vol}$,' indicating a dimension of measurement (volume). Thus, '$\mu_{oz,vol}$' is a function from entities to numbers representing the ratio of entities to a canonical ounce, by volume. These numbers are provided by a numeral, in this case 'four.' Accordingly, 'four ounces' is a predicate true of entities measuring exactly four ounces, cf. (120b). This can then combine with ADJUNCT to produce an

---

[54]   Cf. Chierchia (1998), Scontras (2014), and Rothstein (2017).

intersective modifier,[55] cf. (120c), which in turn can combine with the predicate in (120d). The result is a predicate true of quantities of water measuring exactly four ounces, cf. (120e).

Consequently, (119a) will be true if the deictically indicated quantity consists of water measuring that amount, just as (119b) will be true if the deictically indicated plurality consists of boxes weighing exactly four pounds (in total). Similarly, appealing to A from above, (81h) will be true if there is water measuring four ounces which Mary drank.

Now consider (121), where 'four pounds' occurs as the object of the transitive verb 'weigh.'

(121)   Those boxes weigh four pounds (in total).

Typically, transitive verbs denote relations between entities. For example, 'Mary loves John' is true if Mary stands in the love-relation to John. On its face, then, (121) relates a deictically indicated plurality to a certain weight named by 'four pounds.' Thus, given (120), the question is how 'four pounds' can come to function as a singular term designating that weight.

One simple solution is provided by Scontras (2014)'s analysis of DEGREES, or abstract representation of measurement. Schematically, degrees take the following form, where 'd' ranges over degrees, 'μ_f' is a measure function relativized to a dimension of measurement f, and '$\cap$' is Chierchia (1984, 1998)'s nominalization operation from above.

(122) $d = {}^{\cap}[\lambda x.\, \mu_f(x) = n]$

Thus, according to (122), degrees are special kinds of nominalized properties. Specifically, they are nominalized properties of measured substances, and different sorts of degrees result depending on the dimension of measurement employed. For example, whereas (123a) designates a DEGREE OF VOLUME, (123b) instead designates a DEGREE OF CARDINALITY.

(123)   a. $d = {}^{\cap}[\lambda x.\, \mu_{oz,vol}(x) = 4]$

      b. $d = {}^{\cap}[\lambda x.\, \mu_{\#}(x) = 4]$

At a given world, these different degrees are INSTANTIATED by different substances. For example, (123a) is instantiated by all substances measuring exactly four ounces, while (123b) is instantiated by all pluralities having exactly four atomic parts.

---

[55] For arguments that measure phrases are intersective, see e.g. Schwarzschild (2005) and Rothstein (2017).

With degrees in place, it is easy to see how we can generate a referential meaning appropriate for 'four pounds' in (121): we simply nominalize the predicate 'four pounds' in (124a).

(124)   a. $[\![\text{four pounds}]\!] = \lambda x.\ \mu_{lb,weight}(x) = 4$

       b. $\text{NOM}(\lambda x.\ \mu_{lb,weight}(x) = 4) = {}^{\cap}[\lambda x.\ \mu_{lb,weight}(x) = 4]$

Accordingly, (121) will be true if the deictically indicated boxes stand in the appropriate relation to the degree of weight generated in (124b), namely they *instantiate* it.[56]

A meaning appropriate for 'four' in (81e) can be generated similarly.

(81e)   Jupiter's moons number <u>four</u> (in total).   (verbal complement)

Specifically, we nominalize the corresponding cardinality predicate.

(125) $\text{NOM}(\lambda x.\ \mu_{\#}(x) = 4) = {}^{\cap}[\lambda x.\ \mu_{\#}(x) = 4]$

This is the same degree of cardinality given in (123b), which is subsequently identified with a number on PA. According to the result, (81e) is true if the plurality of Jupiter's moons stands in the appropriate relation to that degree of cardinality, or number, namely the former instantiates the latter.

Incidentally, this is very similar to the analysis of specificational uses I defend in Snyder (2017).

(81d)   The number of Jupiter's moons is <u>four</u>.   (specificational)

As mentioned in §2.3, it has been argued that (81d) is a *specificational* sentence. Suppose so. Now assume the meaning of the specificational copula repeated from §2.3.3 in (76), due to Romero (2005).

(76) $[\![\text{be}]\!] = \lambda x.\lambda \underline{y}_{<s,e>}.\lambda w.\lambda x.\ \underline{y}(w) = x$

As before, '$\underline{y}$' here ranges over individual concepts, i.e. functions from worlds to individuals. Thus, according to (76), the pre-copular material of a specificational sentence designates an individual concept which, at a given world w, is identified with the referent of the post-copular term. For example, (15) will be actually true if, in fact, the unique individual meeting the description 'x composed *Tannhäuser*' is Wagner.

(15)   The composer of *Tannhäuser* is Wagner.

Similarly, I argue in Snyder (2017) that (81d) is actually true if, in fact, the maximal degree of cardinality instantiated by Jupiter's moons is identical to

---

[56] See Scontras (2014) for a similar suggestion.

the one in (123b). This is given in (126), where (126a) provides the meaning of relational, cardinal 'number' (see §2.2.2), and '$\underline{d}$' in (126b) is an individual concept mapping worlds to degrees.[57]

(126) a. $[\![\text{number}_c]\!] = \lambda P.\lambda d.\ d = {}^\cap[\lambda x.\exists n.\ \mu_\#(x) = n \wedge P(x)]$

  b. $\lambda w.\sigma\,\underline{d}[\underline{d}(w) = {}^\cap[\lambda x.\exists n.\mu_\#(x) = n \wedge {}^*\text{JM}(x)]] = {}^\cap[\lambda x.\mu_\#(x) = 4]$

Thus, according to (126a), cardinal 'number' denotes a relation between properties of pluralities and degrees of cardinality those pluralities instantiate. Combining this with specificational 'be' in (76) ultimately results in truth-conditions for (81d) given in (126b): (81d) is actually true if, in fact, the maximal degree of cardinality instantiated by Jupiter's moons is the one generated in (125). In other words, (81d) is actually true if, in fact, Jupiter has *exactly* four moons.

3.2.4 Ordinal Uses

Finally, we turn to ordinal uses of 'four' like (81i).

(81i) Mary is contestant number <u>four</u>.   (ordinal)

(81i) is plausibly equivalent to (127), where 'fourth' is an ordinal adjective.

(127) Mary is the <u>fourth</u> contestant.   (ordinal adjective)

However, unlike with the cardinal and arithmetic uses encountered above, it is not immediately obvious what the semantic function of 'four' in (81i) is. In contrast, the semantic function of the ordinal adjective in (127) is fairly clear: it is a modifier. Thus, the strategy employed here will be to first adopt an independently motivated analysis of ordinal adjectives, and then use that analysis in conjunction with the noted equivalence to glean a seemingly appropriate meaning for 'four' in (81i).

The account of ordinal adjectives adopted here is due to Rothstein (2017).[58] According to it, 'fourth' is morphologically complex, consisting of a numeral 'four' and an ordinal suffix '-th.' The meaning of 'fourth' results from combining these, resulting in a modifier describing the ordinal position of an individual among a linearly ordered class.

Specifically, the ordinal suffix '-th' is given the meaning in (128a), where '| |' is again a cardinality-operation defined on sets.

(128) a. $[\![\text{-th}]\!] = \lambda n.\lambda P.\lambda x.\ P(x) \wedge |\{y : R(x,y)\}| = n - 1$

---

[57] This assumes that degrees are *kinds* in Chierchia (1998)'s sense (cf. Scontras (2014)), and that kinds are *individuals* in Link (1983)'s sense (cf. Chierchia (1998)).

[58] Though see Bylinina et al. (2014) for a similar account.

b. $\llbracket \text{four} \rrbracket = 4$

c. $\llbracket \text{-th} \rrbracket (\llbracket \text{four} \rrbracket) = \llbracket \text{fourth} \rrbracket = \lambda P.\lambda x.\ P(x) \wedge |\{y : R(x,y)\}| = 4 - 1$

d. $\llbracket \text{fourth contestant} \rrbracket = \lambda x.\ \text{con}(x) \wedge |\{y : R(x,y)\}| = 3$

e. $\llbracket \text{the fourth contestant} \rrbracket = \sigma x[\text{con}(x) \wedge |\{y : R(x,y)\}| = 3]$

As before, free 'R' indicates that its meaning is determined by context. In the present case, it could mean e.g. 'x is preceded by y in performing in the competition.' This imposes a linear order on a salient set of competitors, and the job of the numeral 'n' in (128a) is to supply the cardinality of that set. Thus, combining '-th' with the numeral 'four' in (128b) results in the modifier in (128c), or the meaning of 'fourth,' which can then combine with a noun like 'contestant' to return a predicate true of contestants such that exactly three things precede them in the competition, cf. (128c). Finally, assuming the same meaning of 'the' from above, 'the fourth contestant' will thus be a singular term referring to a uniquely salient such contestant. Consequently, (127) will be true if Mary is that contestant.

Turning now to (81i), the question becomes: How can we generate a plausible meaning for 'contestant number four' which renders (81i) and (127) equivalent? Like (127), I assume (81i) is an identity statement, so that 'contestant number four,' like 'the fourth contestant,' functions as a singular term. What's needed, then, is a meaning similar to that contributed by the ordinal suffix '-th' in (128a). There would appear to be two options: it is contributed by 'four,' or else by (ordinal) 'number.' But since 'four' functions as a numeral in (127), the natural assumption is that it does likewise in (81i), thus providing a numerical argument to (ordinal) 'number.'

Thus, suppose ordinal 'number' has the meaning given in (129a).

(129)   a. $\llbracket \text{number}_o \rrbracket = \lambda n.\lambda P.\lambda x.\ P(x) \wedge |\{y : R(x,y)\}| = n - 1$

b. $\llbracket \text{number}_o \text{ four} \rrbracket = \lambda P.\lambda x.\ P(x) \wedge |\{y : R(x,y)\}| = 4 - 1$

c. $\llbracket \text{contestant number}_o \text{ four} \rrbracket = \lambda x.\ \text{con}(x) \wedge |\{y : R(x,y)\}| = 3$

According to (129a), ordinal 'number' is synonymous with the ordinal suffix '-th.' So, combining the former with the numeral 'four' and the noun 'contestant' results in the same predicate in (128d). This can then combine with IOTA to generate a singular term coreferential with the one in (128e). (81i) and (127) are thus predicted to be equivalent: both will be true if Mary is the unique contestant preceded by exactly three others in the competition.

3.2.5 Summary of the Semantics

Summarizing, there are broadly speaking two kinds of polymorphic analyses available. According to PS, the lexical meaning of 'four' is that of a numeral

Example	Semantic Type	Denotation
(81f), (81h), (81i)	n	4
(81g)	$\langle n, t \rangle$	$\lambda n.\, n = 4$
(81a)	$\langle e, t \rangle$	$\lambda x.\, \mu_\#(x) = 4$
(81d), (81e)	d or n	$^\cap[\lambda x.\, \mu_\#(x) = 4]$
(81b)	$\langle \langle e, t \rangle, \langle e, t \rangle \rangle$	$\lambda Q.\lambda x.\, \mu_\#(x) = 4 \wedge Q(x)$
(81c)	$\langle \langle e, t \rangle, \langle \langle e, t \rangle, t \rangle \rangle$	$\lambda P.\lambda Q.\exists x.\, \mu_\#(x) = 4 \wedge P(x) \wedge Q(x)$

naming the number four, while according to PA, it is instead that of a predicative adjective counting the atomic parts of a plurality.

(99a) $[\![\text{four}]\!] = 4$ (PS)

(99b) $[\![\text{four}]\!] = \lambda x.\, \mu_\#(x) = 4$ (PA)

Given either lexical meaning, it is possible to derive meanings appropriate for the various occurrences of 'four' in (81) via independently motivated and well-attested type-shifting principles, as summarized in table above.

The derivations provided are summarized by the following map, where arrows indicate directionality, and "RSE" abbreviates Rothstein (2013, 2017)'s schematic equation, repeated in (116):

(116) $n = {}^\cap[\lambda x.\, \mu_\#(x) = n]$

There are two important observations here. First, on both PS and PA, 'four' has genuinely referential uses, whose referent may be a number or a nominalized cardinality property, if those are different. Regardless, independent of its *lexical* meaning, 'four' functions as a singular term in some syntactic contexts, referring to a number. Secondly, there is an element shared by all potential meanings of 'four': the number four. In a sense, it serves as the formal witness to the polysemy of 'four,' tying all of its potential meanings together. Both observations will play a critical role in what follows.

3.3 An Improved Argument for Realism

So far, we have seen that number expressions are polymorphic, and that two kinds of polymorphic analyses are available. Both not only provide meanings appropriate for all of the various uses of number words noted, but also explain how those meanings are systematically related. Because these exhaust the relevant options – the lexical meaning of e.g. 'four' is either referential in character or else non-referential in character – I will assume hereon that one of them is correct. The question I want to pursue here is what ramifications this may have for the ontology of number.

Recall that Hale's Argument consists of three theses, including Singular Terms and Candidacy. Both claims can be reasonably doubted. On the neo-Fregean paradigm of Hale (1987, 2016) and Hale and Wright (2001), ontological questions are to be decided wholly on the basis of broadly semantic considerations. Because singular terms characteristically denote objects, the question of what objects there are is subsumed under the question of what singular terms feature in true statements. What's required, then, is an *independent* characterization of singular terms, one which does not beg ontological questions in favor of neo-Fregean realism.

On the other hand, finding such a characterization has proven exceedingly difficult. For example, it will not do, apparently, to define 'singular term' by reference to surface grammar, as Hale (2016, p. 15–16) explains:

> It is easy to see that we cannot satisfactorily circumscribe the class of singular terms purely by reference to surface grammar, say as comprising just singular nouns or noun-phrases. For many expressions which we should rightly refuse to count as singular terms—including, importantly, words serving to express generality such as 'everything,' 'nothing,' 'something,' along with restricted quantifier phrases, such as 'every philosopher,' 'no policeman,' 'some city,' etc.—may occupy, without violence to grammar or sense, positions in sentences in which simple proper names, or other paradigm examples of singular terms, may stand.

Within quantificational logic, the underlined expressions in (130) are analyzed differently: referentially and quantificationally, respectively.

(130)   {Mary/Nothing} danced.

Yet both expressions serve the same grammatical role as subjects. So, surface grammar alone will not justify a divergence in analysis.

Hence the motivation for Dummett (1973)'s *inferential* characterization, whereby singular terms are exactly those expressions licensing the inferences indicated in Section 2. However, Dummett's characterization is inadequate, according to Hale (1987, 1994, 2016), because it admits counterexamples. In response, Hale has devised his own inferential characterization, one which is significantly more complex (and arcane) than Dummett's. However, it too has been accused of admitting counterexamples.[59] Suffice it to say that, at present, it is not clear whether 'singular term' can be characterized in a way suitable for neo-Fregean purposes. Yet without some such characterization in place, it is hard to see how Singular Terms could be justified.

Candidacy is also potentially problematic. Following Frege, neo-Fregeans assume a particular conception of natural numbers, which serve as the referents of number expressions used referentially. Specifically, natural numbers are *finite cardinals* generated by HP (see §2.2.1). Yet as mentioned in §2.2.2, there are good empirical reasons for thinking that natural language distinguishes numbers and cardinalities, and that not all apparently referential uses refer to the latter. Thus, it is far from obvious that, *in general*, the best candidate referents of such uses are "numbers" in Hale's sense.

The purpose of this section is to shore up these potential weaknesses in Hale's Argument, by placing Singular Terms and Candidacy on firmer empirical footings, thus resulting in an improved argument for realism. Specifically, I claim that polymorphic analyses afford stronger arguments for both theses, suitably understood, beginning with Singular Terms.

3.3.1 The Empirical Case for Singular Terms

If singular terms are not to be characterized inferentially, then how else might they be? Within contemporary semantic theory, the answer is: *type-theoretically*. Specifically, singular terms are those expressions whose semantic type is *e*, denoting an entity.[60] Of course, polymorphism complicates matters somewhat, allowing some expressions to take on different semantic types in different syntactic contexts. As a result, the semantic type of an expression cannot generally be determined independent of the syntactic context in which it occurs. For this reason, it is better to characterize singular terms, and the semantic function of an expression more generally, in terms of *occurrences*. Specifically, I propose the following definition: an occurrence of an expression is a SINGULAR TERM just in case its semantic type, in the syntactic context of that occurrence, is type-*e*.

Given this characterization, the question of whether we should accept Singular Terms becomes: Are there occurrences of number expressions in true numerical statements such that the semantic type of that expression, in the context of that statement, is type-*e*? This is an empirical question, of course, one

---

[59] See e.g. Schwartzkopff (2016).
[60] See e.g. Partee (1986b).

which presumably depends on the best available semantic evidence. And on all extant polymorphic analyses, the answer is clearly "Yes." So, if polymorphic analyses are in fact our best available semantic theories, as I have argued, then Singular Terms is correct.

But *why* do extant polymorphic analyses analyze certain occurrences of number expressions as singular terms, in this sense? There are different reasons. One appeals to compositionality. Note that given orthodox semantic assumptions, the underlined expressions in (131) are type-*e*.

(131) a. <u>Mars</u> is red.

   b. <u>The planet Mars</u> is red.

   c. [Pointing at Mars:] <u>That planet</u> is red.

There are independent semantic reasons for assigning (occurrences of) names like 'Mars' in (131a) type-*e*. For example, as mentioned by Partee (1986b), unlike e.g. 'nothing,' 'Mars' licenses the kind of anaphora witnessed in (132).

(132) {<u>Mars</u>/??<u>Nothing</u>} is mysterious. <u>It</u> is also what I'm writing about.

The intuitive equivalence of (131a-c) provides another argument. Specifically, assuming 'Mars' is type-*e*, employing the same type-shifting principles from §3.2.2 and combining them with 'Mars' and 'the' produces (133), i.e. a singular term coreferential with 'Mars.'

(133) ⟦the planet Mars⟧ $= \sigma y[\texttt{planet}(y) \wedge y = m]$

Note that (133) is possible only if 'Mars' in (131b) is shifted *from* type-*e* to type-$\langle e, t \rangle$. Put differently, if 'Mars' in (131a) were not type-*e*, then type-shifting would fail to generate (133). Finally, combining an independently motivated meaning for 'that' with 'planet' would result in another coreferential singular term.[61] However, the resulting truth-conditions are equivalent to those for (131a,b) only if the entity deictically referenced by 'that planet' is the same one referenced by 'Mars' in (131a) and 'the planet Mars' in (131b). Thus, compositionally explaining the equivalence of (131a-c) requires recognizing that all three of the underlined expressions are coreferential singular terms, in the above sense.

Now, we saw in §3.2 that on both kinds of polymorphic analyses, the parallel underlined expressions in (134) are also type-*e*.

(134) a. <u>Four</u> is even.

   b. <u>The number four</u> is even.

   c. [Pointing at the numeral 'four':] <u>That number</u> is even.

---

[61] See Scontras (2014) and Snyder (2020).

Minimally, compositionality requires a mapping from syntax to semantic representation. This imposes certain kinds of *uniformity constraints*. For example, bare names in argument positions, such as 'Mars' in (131a), should have a uniform semantic type, compelling evidence to the contrary notwithstanding. Since 'four' in (134a) is a name by any reasonable syntactic criterion,[62] and since it too gives rise to anaphora (§3.2.2), 'four' in (134a) should also be type-*e*. Similarly, meanings appropriate for 'the number four' and 'that number' should arise in the same way as those for 'the planet Mars' and 'that planet.' Thus, not only should 'four' in (134b) get shifted from type-*e*, 'that number' in (134c) should be a singular term coreferential with the numeral 'four' and 'the number four,' since (134a-c) are equivalent.

A similar kind of argument is provided by predicative and specificational uses, such as those in (135a,b).

(135)   a.   Jupiter's moons are four (in number).

   b.   The number of Jupiter's moons is four.

As noted, color words and measure phrases have similar uses.

(136)   a.   Jupiter's moons are green (in hue).

   b.   The color of Jupiter's moons is green.

Clearly, the meanings of the copulas involved in these pairs are different. For example, unlike with (136a), (136b) cannot be reasonably understood as claiming that a certain *color* is green. Rather, the b-sentences are specifying a particular cardinality or color instantiated by the subjects of the a-sentences. Despite this, the pairs given are plausibly equivalent: (135a,b) are both true if Jupiter has exactly four moons, for instance. What we want, ideally, is a *uniform* analysis which not only predicts these equivalences, but which also explains how the predicates involved are semantically related.

Again, the analysis sketched in §3.2.3 does just that. According to it, the a-sentences are true just in case the entities referenced by the subject have the property predicated. Equivalently, the b-sentences are true just in case the nominalized properties specified by the copula are instantiated by those same entities. Thus, the semantic relationship between the two copulas is evident: whereas those in the a-sentences involve predicating a property, those in the b-sentences involve naming that same property. Crucially, though, this explanation holds only if 'four' in (135b), like 'green' in (136b), functions as a *singular term*, in the above sense.

---

[62] Cf. Moltmann (2013a,b).

A final, similar argument appeals to intuitive equivalences between predicative and verbal complement uses, such as (137a,b) and (138a,b).

(137) a. Jupiter's moons are <u>four in number</u>.

      b. Jupiter's moons <u>number four</u>.

(138) a. Those boxes are <u>four pounds in weight</u>.

      b. Those boxes <u>weigh four pounds</u>.

Again, what we want is a compositional analysis which not only predicts these equivalences, but which also explains the clear relationship in meaning between the underlined expressions involved. And, again, the analysis developed in §3.2.3 affords just that. Specifically, assume the meanings in (139a,b), where 'CARD' and 'WEIGHT' denote sets of degrees of cardinality and weight, respectively, and '$\cup$' is Chierchia (1998)'s predicativizing-operation, which cancels the converse nominalizing-operation $\cap$.

(139) a. $[\![\text{in number}]\!] = \lambda d \in CARD.\lambda x.\ ^{\cup}d(x)$

      b. $[\![\text{in weight}]\!] = \lambda d \in WEIGHT.\lambda x.\ ^{\cup}d(x)$

Notably, these are the same meanings assumed for the verbs 'number' and 'weigh' in §3.2.3. Now, nominalizing 'four' and 'four pounds' in the manner suggested there and combining the result with (139a,b) delivers:

(140) a. $[\![\text{four in number}]\!] = \lambda x.\ ^{\cup\cap}[\lambda y.\ \#(y) = 4](x)$

      b. $[\![\text{four pounds in weight}]\!] = \lambda x.\ ^{\cup\cap}[\lambda y.\ \mu_{lb,vol}(y) = 4](x)$

These are equivalent to the denotations assumed for predicative 'four' and 'four pounds' above. Thus, as desired, (137a,b) and (138a,b) are predicted to be equivalent: the latter are both true if those boxes weigh exactly four pounds, for instance. Again, however, this explanation succeeds only if 'four' in (137a,b) functions as a *singular term*.

A second kind of empirical argument for Singular Terms is provided by an important distinction drawn throughout natural languages between counting and measuring. In English, this is made evident in so-called CONTAINER PHRASES like "glass of water," which are known to be ambiguous between INDIVIDUATING INTERPRETATIONS (IIs) and MEASURE INTERPRETATIONS (MIs), illustrated respectively in (141b,c).[63]

I-Context: Mary has a strange way of heating water for coffee. She fills four glasses with water and sets those glasses into the boiling soup. Pointing at the soup, John says:

---

[63] Cf. Landman (2004), Rothstein (2010, 2017), Partee and Borschev (2012), Scontras (2014), Snyder and Barlew (2016, 2019), and Snyder (2020).

M-Context: Mary wants to make soup. The recipe calls for four glassfuls of water. Estimating, she pours water directly from the tap into the soup. Pointing at the soup, John says:

(141) a. Mary put four glasses of water in the soup.

b. Mary put four glasses filled with water in the soup. (II)

c. Mary put four glassfuls of water in the soup. (MI)

Clearly, (141a) receives different interpretations in these different contexts. In particular, it receives an II in the I-Context, and a MI in the M-Context. That's because we are doing different things with 'four' in these different contexts. Specifically, in the I-Context, we are using it to *count* four glasses, each of which happens to be filled with water. In the M-Context, however, we are using it to *measure* an amount of water, namely the amount which would fill a certain glass four times.

The philosophical significance of I/M ambiguities is three-fold. First, they suggest that there is a linguistically significant distinction between counting and measuring, a thesis which arguably traces back to Frege (1903).[64] Secondly, they suggest that not all seemingly prototypical *non*-referential uses of 'four' can be analyzed quantificationally. Specifically, only (141b) is plausibly analyzed as (142), where '*G*' represents 'glass of water.'

(142) $\exists x_1, ..., x_4. [G(x_1) \wedge ... \wedge G(x_4) \wedge x_1 \neq x_2 \wedge ... \wedge x_3 \neq x_4] \wedge [\forall z. G(z) \rightarrow z = x_1 \vee ... \vee z = x_4]$

This corresponds to the II, of course, where we are *counting* glasses of water as individuated entities. Thus, if 'four' in (141a) *only* had this kind of meaning, there would be no ambiguity to account for.

Finally, and in light of this, standard analyses explain the distinction between IIs and MIs in terms of the semantic function of 'four.' Specifically, the II arises if 'four' in (141a) functions as a cardinal modifier, counting glasses filled with water, while the MI arises if instead 'four' functions as a *singular term*, providing the numerical argument to a measure function denoted by 'glasses of water,' just like with other measure nouns (§3.2.3).[65] If so, then I/M ambiguities, and the counting/measuring distinction more generally, provide independent empirical support for Singular Terms. That is, independent of whether *explicitly* referential uses like (134a-c) are type-*e*, making semantic

---

[64] See Snyder and Shapiro (2016). To be clear, I/M ambiguities are not the only linguistic evidence for distinguishing counting from measuring. Indeed, there is plenty of additional, cross-linguistic evidence. See especially Scontras (2014) and Rothstein (2017).

[65] See the references in fn. 63.

sense of the counting/measuring distinction plausibly requires recognizing that certain occurrences of 'four,' for instance, are.

In sum, there is overwhelming empirical evidence that number expressions are singular terms in many true numerical statements. It bears emphasizing, however, that the resulting motivation for Singular Terms, suitably understood, is not vague intuitions about underlying logical form, surface grammatical role, or some pre-theoretic notion of singular termhood characterized inferentially, assuming such a notion exists.[66] Rather, it is providing a comprehensive, empirically adequate polymorphic semantics, something which is independently theoretically desirable. As we have seen, constructing such a semantics will almost certainly require recognizing referential occurrences of number expressions, contra non-referentialism.

3.3.2 The Empirical Case for Candidacy

We saw in §3.2.5 that on standard polymorphic analyses, the various meanings number expressions can take on are related in virtue of sharing a certain element in common, namely a number. Put differently, numbers are, in a sense, the formal witness to the polysemy of number words. This provides a new, distinctly empirical basis for Candidacy, suitably qualified.

Recall the map of potential meanings from §3.2.5.

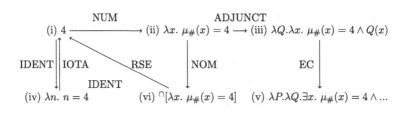

Whereas the lexical meaning of 'four' is located at node (i) on PS, it is instead located at node (ii) on PA. On both analyses, however, numbers form the referents of numerals featuring in arithmetic statements, measure phrases, and ordinal uses. They also serve as values of measure functions like $\mu_\#$, common to the meanings of all cardinal uses.

Now, suppose for argument's sake that numbers do not exist. Then *none* of the potential meanings indicated in the map would be available. Specifically, since the lexical meaning of 'four' on PS would not be available, there would be no input to the type-shifting principles employed to generate subsequent meanings appropriate for other uses. Similarly, because there would be

---

[66] Especially since singular termhood is, or should be, considered a *technical* notion.

no measure functions, the lexical meaning of 'four' on PA would not be available, and so no subsequent meanings could be generated. A fortiori, on neither analysis could number expressions have *different* meanings, let alone different but *related* meanings. In other words, if numbers did not exist, then number words would not be *polysemous*, contrary to fact.

Simply put, our best available explanation for why number words are polysemous presupposes realism. But *why* do extant polymorphic analyses employ numbers in this manner? After all, couldn't some other kind of entity potentially play the role of numbers within those analyses?

The answer, it seems, appeals to general considerations of theoretical economy. First, given the truth of (81f), numbers are minimally needed to serve as the referents of numerals figuring in arithmetic statements.

(81f) Four is an even number.

Furthermore, measure functions are independently needed to capture the meanings of measure phrases, for reasons broached in §3.3.1. Thus, general considerations of theoretical economy recommend identifying the referents of numerals featuring in arithmetic statements with those featuring in measure phrases, all else being equal. Furthermore, once we recognize that meanings appropriate for cardinal uses can also be obtained via measure functions, and that meanings of ordinal adjectives are complex and plausibly exploit counting, considerations of theoretical economy recommend positing a *single* domain of entities, which *all* of these meanings exploit.[67]

Ultimately, then, there appear to be excellent *empirical* reasons for endorsing Candidacy. Numbers are not merely the *intuitive* referents of number words used referentially in arithmetic equations or statements of number theory. Rather, explaining the *polysemy* of number words requires recognizing that their various meanings have an element in common, one which, for reasons of economy, is witnessed in referential *and* non-referential uses alike.

It need not follow from this, however, that numbers are the most plausible candidate referents of *all* referential occurrences of number words. Indeed, we have seen that Candidacy may not hold in full generality, for strictly empirical reasons. Recall contrasts such as (27a,b), motivating a distinction between cardinalities and what Moltmann calls "pure numbers."

(27a) The number of women is {four/??the number four}.

(27b) The number Mary is thinking about is {four/the number four}.

---

[67] Of course, polymorphic analyses do not alone determine the *nature* of the numbers posited: whether they are abstract, or whether they should be identified with certain sets, equivalence classes of equinumerous concepts, positions within ω-sequences, etc. Ultimately, these are matters of METASEMANTICS, or metaphysics, not semantics proper, and addressing them adequately is far beyond the scope of the present monograph.

As mentioned in §2.2.2, such contrasts cast doubt on the neo-Fregean identification of numbers with cardinalities, or equivalence classes of equinumerous concepts. Now recall Scontras (2014)'s schema for degrees.

(122) $d = \,^{\cap}[\lambda x.\, \mu_f(x) = n]$

Here, degrees and numbers are different *sorts* of entities. Specifically, degrees are nominalized properties of measured substances, where the notion of measurement involved presupposes an *independent* domain of numbers, serving as values of measure functions.

I argue in Snyder (2017) that adopting this conception of degrees affords a plausible explanation of contrasts like (27a,b). Specifically, in keeping with §3.2.3, whereas 'the number of women' references a degree of cardinality, 'the number Mary is thinking about' instead references a number. Moreover, since 'the number four' refers to a number, it is the same sort of entity as referenced by the pre-copular material in (27b), but not in (27a), thus explaining the contrast. Crucially, however, this explanation goes through only if the sorts of things answering 'how many'-questions – degrees of cardinality – and numbers are distinct. If so, then Candidacy should be qualified: numbers are the most plausible candidate referents for number expressions used referentially in many, but not necessarily *all*, cases. In cases like (27a), for instance, they instead refer to degrees.

On the other hand, this requires taking a stance on the two kinds of polymorphic analyses sketched above. As mentioned, Rothstein (2013, 2017) stipulates the schematic equation in (116) to ensure that the individual property correlates generated by cardinal nominalization are the same sorts of things forming the extension of arithmetic 'number.'

(116) $n = \,^{\cap}[\lambda x.\, \mu_{\#}(x) = n]$

Obviously, (116) is inconsistent with the hypothesis that natural language sortally distinguishes numbers and nominalized cardinality properties. So, if (116) is correct, then Candidacy may hold in full generality after all.

In short, whether Candidacy holds in full generality may well depend on what the best explanation of semantic contrasts like those noted in §2.2.2 turns out to be. This is an empirical matter, of course. Regardless, since it suffices to establish realism that at least *some* referential uses of number expressions refer to numbers, it is not essential to Hale's Argument that Candidacy holds in full generality. Moreover, since all available polymorphic analyses substantiate a suitably qualified form of Candidacy, the best available semantic evidence ultimately supports realism.

3.3.3 Consequences for the Easy Argument

I have argued that polymorphic analyses provide novel empirical support to Singular Terms and Candidacy, suitably understood. Another question to be addressed is whether the semantics developed in §3.2 can shed any light on the philosophical dialectic sketched in Section 2. Interestingly, although both polymorphic analyses sketched above support realism, they may not both straightforwardly vindicate the Easy Argument.

The crucial contention, recall, was that Frege's (11b) is an identity statement. If so, then post-copular 'four' is a singular term. Hence, (11b) would entail (17), which apparently quantifies over numbers.

(17)   There is a number which is the exact number of Jupiter's moons, namely four.

We then saw that on some versions of adjectivalism, (11b) is not a genuine identity statement, strictly speaking, but rather a *specificational* sentence. If so, then the potential problem with the Fregean analysis is that it relies too much on surface syntactic appearances: even though (11b) looks like it equates the referents of singular terms, in reality it does not.

Now, suppose the analysis of (11b) proposed in §3.2.3 is correct. According to it, recall, combining the meaning of the specificational copula with the meaning of cardinal 'number' ultimately results in truth-conditions provided in (126b): (11b) is actually true if, in fact, the maximal degree of cardinality instantiated by Jupiter's moons is the nominalized cardinality property of being four in number.

(126b) $\lambda w. \sigma \underline{d}[\underline{d}(w)] = \cap[\lambda x.\exists n. \mu_\#(x) = n \wedge JM(x)]] = \cap[\lambda x. \mu_\#(x) = 4]$

What's more, a distinguishing feature of PA is that it identifies precisely the sorts of nominalized cardinality properties referenced in (126b) with *numbers*. Hence, the Easy Argument would appear straightforwardly valid, given present assumptions: post-copular 'four' would be a nominalized adjective referring directly to a number. Thus, even if (11b) is a specificational sentence, it may straightforwardly entail (17), thereby vindicating realism.

On the other hand, since numbers are *semantic primitives* on PS, they may or may not be identified with nominalized cardinality properties. Moreover, as mentioned, contrasts like those in §2.2.2 arguably motivate distinguishing these. If so, then the problem with the Easy Argument is that it *equivocates*: it conflates what natural language distinguishes, namely arithmetic 'number' and cardinal 'number,' and thus numbers and cardinalities. Or so I argued in Snyder (2017), at least.

Ultimately, I believe that this nicely illustrates some of the more pressing empirical challenges facing *both* kinds of polymorphic analyses. The clearest empirical advantage of PA is that it, unlike PS, generates meanings for numerals similarly to other nominalized adjectives, such as 'green.'

(88) Green is a secondary color.

In both cases, an appropriate meaning is generated via nominalization. On the other hand, it is far from clear whether an analog of PS would work for color words. Suppose, for instance, that 'green' in (88) names a certain color, understood as a primitive atomic entity.[68] Clearly, the type-shifting principle responsible for generating predicative uses of 'four' on PS, namely NUM, is not applicable to such an entity.

(102b)　$\text{NUM} = \lambda n.\lambda x.\ \mu_{\#}(x) = n$

Thus, it is not at all obvious how meanings for non-referential uses of 'green' should be obtained on this analysis.

On the other hand, it is also unclear whether PA succeeds in generating meanings appropriate for numerals via nominalization. That's because the numbers featuring in the schematic equation (116) are the very *same* entities serving as arguments to measure functions like $\mu_{\#}$. In other words, as a kind of definition, (116) appears *viciously circular*. To see why, consider that true identities generally license substitution of like for like. Yet substituting a nominalized cardinality property for the numerical argument of '$\mu_{\#}$' in (116) leads to a never ending chain of subsequent nominalizations:

(143)　$n = {}^{\cap}[\lambda x.\ \mu_{\#}(x) = {}^{\cap}[\lambda x.\ \mu_{\#}(x) = {}^{\cap}[\lambda x.\ \mu_{\#}(x) = \ ...$

Simply put, numbers are needed *prior to* nominalization, and yet nominalization is supposed to *generate* entities which are numbers.

This is not the place to settle these matters. Minimally, what this shows is that more work is needed for both kinds of polymorphic analyses. Indeed, the question of which kind of polymorphic analysis is ultimately correct is far from settled. Likewise for the status of the Easy Argument, it seems. Regardless, I have argued here that the case for realism does not stand or fall with the Easy Argument. Independent of which kind of polymorphic analysis is ultimately correct, certain occurrences of number expressions will need to function referentially in true number-related statements, and numbers will be needed to explain their polysemy.

---

[68] As with analyses such as Kennedy and McNally (2005).

3.4 Conclusion

I began this monograph with a core problem within contemporary philosophy of mathematics, namely Benacerraf's Dilemma. The key question was whether we should accept the semantic horn: that number words featuring in true arithmetic statements refer to numbers. To this end, I introduced Hale's Argument, which spells out that horn. The contested premise is Singular Terms: number expressions function as singular terms in a variety of true mathematical statements, broadly construed. If so, then it would appear that their truth commits us to realism. On the other hand, if non-referentialism is correct, and so all apparently referential uses of such expressions are in fact non-referential, then the truth of our number talk is not only consistent with nominalism, it may ultimately support it.

I illustrated this debate via an instance of Hale's Argument – the Easy Argument for Numbers – along with two popular philosophical strategies designed to support or undermine it. Whereas substantivalism suggests that seemingly prototypical non-referential uses of number words are in fact referential, thus supporting realism, adjectivalism suggests that seemingly prototypical referential uses are really non-referential, consistent with nominalism. On the other hand, we saw that extant versions of both strategies face significant empirical and theoretical challenges.

In the present section, I have argued that both strategies are ultimately flawed. Not only do number words have a wide variety of *both* referential and non-referential uses, thus resolving Frege's Other Puzzle, explaining how the meanings of these various uses are related ultimately supports referentialism, and indeed realism, but for reasons which have been largely unappreciated within the philosophical literature. Specifically, explaining the polysemy of number words requires adopting a polymorphic analysis, and all extant analyses not only recognize genuinely referential uses of number words, they also posit numbers as the entities relating their numerous potential meanings, for independent and empirically motivated reasons.

How might nominalists respond to the resulting argument for realism? I will consider three options. The first would be to construct an alternative polymorphic analysis, one which not only supplies empirically adequate meanings for all uses of number words in a compositional manner, but uses only independently motivated and well-attested type-shifting principles to do so, all without using numbers. While I cannot see how to rule out this possibility, in principle, it does strike me as an exceedingly difficult task.

The second nominalist alternative would be to adopt some form of error theory, also sometimes called FICTIONALISM. Traditionally, error theory is the conjunction of two theses. First, arithmetic statements (at least), such as '2 + 2 = 4,' should be taken at face-value, so that apparent singular terms featuring in those statements are in fact singular terms. Hence, '2 + 2 = 4' is true just in case the referent of '4' is the same number as the referent of '2 + 2.' Secondly, nominalism is correct, so that the would-be referents of these singular terms do not exist. Consequently, arithmetic statements (at least) generally either suffer from presupposition failure or else are straightforwardly false. In either case, they are not true.

Again, I cannot see how to rule out this possibility, in principle. However, it is noteworthy that if the semantics sketched above is correct, then adopting error theory in full generality requires attributing far more error than has been traditionally recognized. Specifically, recalling §3.3.2, *none* of the potential meanings of 'four' discussed above would be available if numbers do not exist, including, but not limited to, meanings appropriate for arithmetic equations. Thus, *none* of the number-related things we say would reflect how the world really is if nominalism were correct. Yet surely *some* of the number-related things we say are true, e.g. that Mars has two moons, or that Michael Jordan is almost two meters tall.

A final nominalist gambit would be to adopt some form of what I call THE PARAPHRASTIC STRATEGY. Although there are many versions available, the unifying theme is that we should not take arithmetic statements at face-value, but rather we should paraphrase, translate, construe, or analyze them so as to render their truth consistent with nominalism. For example, Hodes (1984), Hellman (1989), and Chihara (1990) all suggest analyzing '2 + 2 = 4' as very roughly equivalent to (144).

(144)   Necessarily, if numbers exist, then 2 + 2 = 4.

Clearly, the truth of (144) does not require the actual existence of numbers. So, if '2 + 2 = 4' is equivalent to (144), in some sense of "equivalent," then the truth of the former is consistent with nominalism.

The familiar complaint with the paraphrastic strategy is that it is not at all clear what the suggested paraphrase or analysis is intended to represent. Indeed, Burgess and Rosen (1997) formulate a dilemma based on the two most obvious candidates. On the HERMENEUTIC interpretation, (144) represents a *semantic* hypothesis: it gives the actual meaning of '2 + 2 = 4.' On the REVOLUTIONARY interpretation, (144) represents a *normative* thesis: we should change mathematics so as to render it consistent with nominalism. The apparent dilemma is that if the hermeneutic interpretation is intended, then nominalists should submit their analysis to a linguistics journal; if the revolutionary interpretation is intended, then it should be submitted to a mathematics journal; but since nominalists will be unwilling to do either, the analysis proposed is ultimately fruitless.

Whether this actually constitutes a dilemma for the paraphrastic strategy is unimportant for present purposes. What is important is that unless the hermeneutic interpretation is intended, the paraphrastic strategy does not engage with Hale's Argument. Obviously, as an empirical thesis, referentialism is consistent with the claim that mathematical statements *should* be interpreted (counterfactually) in a manner consistent with nominalism. In fact, it is hard to see why such a normative claim would be made without *presupposing* that arithmetic talk functions in the manner referentialists claim. After all, if non-referentialism were correct, then the descriptive facts alone would suffice to defend nominalism. Yet if number words function referentially in arithmetic statements and nominalism is correct, the result is a form of *error theory*: '2 + 2 = 4,' for instance, is not true.

On the other hand, if the hermeneutic interpretation is intended, then Hale's Argument potentially collapses. Indeed, if (144) provides the actual meaning of '2 + 2 = 4,' and if similar nominalist paraphrases could be found for other apparently referential uses of number words, then Singular Terms could be false. Of course, in that case, nominalist paraphrases would need to be viewed as *competitors* to the polymorphic analyses sketched in §3.2, and thus our best extant semantic theories. And, again, I cannot see how such a proposal could be ruled out, in principle.

However, advocates of the paraphrastic strategy typically reject the hermeneutic interpretation: the paraphrases given are not intended to provide the actual meanings of the sentences paraphrased.[69] And for good reasons. For one, there is no pretense that the paraphrases provided arise *compositionally*. Specifically, it is not assumed that component meanings of '2 + 2 = 4' combine so as to result in truth-conditions suggested by (144). Furthermore, there is little empirical plausibility to the claim that '2 + 2 = 4' and (144) are *synonymous*. For example, whereas the former plausibly entails that some number is equal to 2 + 2, the latter does not.

I conclude that the paraphrastic strategy most plausibly collapses into error theory. If so, then the most familiar nominalist strategies – non-referentialism, error theory, and the paraphrastic strategy – are ineffective responses to the semantic horn of Benacerraf's Dilemma. In this respect, I agree with David Lewis (1986, p. 108–109), who says:

> I think it is very plain which horn of Benacerraf's dilemma to prefer. To serve epistemology by giving mathematics some devious semantics would be to

---

[69] See e.g. Hellman (1998) and Chihara (2004).

reform mathematics. Even if verbal agreement with mathematics as we know it could be secured—and that is doubtful—the plan would be to understand those words in a new and different way. It's too bad for epistemologists if mathematics in its present form baffles them, but it would be hubris to take that as any reason to reform mathematics. Neither should we take that as any reason to dismiss mathematics as mere fiction; not even if we go on to praise it as very useful fiction, as in Hartry Field's instrumentalism. Our knowledge of mathematics is ever so much more secure than our knowledge of the epistemology that seeks to cast doubt on mathematics.

Of course, if the semantics developed here is a reliable guide, then the epistemological horn of Benacerraf's Dilemma would appear far more challenging than its original formulation suggests. Specifically, if numbers are abstracta featuring in the contents of our number-related beliefs, so that those beliefs represent abstracta, and yet representing requires a causal connection between agents and what is represented, then how could *any* of our number-related beliefs be true, let alone those whose contents are specific to mathematics? Lewis is right, I think, that our knowledge of mathematics is far more secure than that of causal epistemologies, and we presumably want to say the same about more mundane number-related beliefs, e.g. that Mars has two moons or that Adams was the second U.S. President. Yet how we are even capable of *entertaining* these beliefs remains a genuine mystery, one which is better suited for a different, and much longer, monograph.

Bibliography

Alexiadou, A. (2017). Deriving color adjectival nominalizations. *Linguística: Revista de Estudos Linguísticos da Universidade do Porto*, 8:143–158.

Balcerak-Jackson, B. (2013). Defusing easy arguments for numbers. *Linguistics and Philosophy*, 36:447–461.

Barker, C. (1998). Partitives, double genitives and anti-uniqueness. *Natural Language and Linguistic Theory*, 4:679–717.

Barwise, J. and Cooper, R. (1981). Generalized quantifiers and natural language. *Linguistics and Philosophy*, 1:413–458.

Benacerraf, P. (1965). What numbers could not be. *The Philosophical Review*, 74(1):47–73.

Benacerraf, P. (1973). Mathematical truth. *Journal of Philosophy*, 70(19): 661–679.

Boolos, G. (1985). Nominalist platonism. *The Philosophical Review*, 94(3): 327–344.

Breheny, R. (2008). A new look at the semantics and pragmatics of numerically quantified noun phrases. *Journal of Semantics*, 25(2):93–139.

Brogaard, B. (2007). Number words and ontological commitment. *Philosophical Quarterly*, 57:1–20.

Burgess, J. P. and Rosen, G. (1997). *A Subject With No Object: Strategies for Nominalistic Interpretation of Mathematics*. Oxford University Press.

Bylinina, L., Ivlieva, N., Podobryeav, A., and Sudo, Y. (2014). A non-superlative semantics for ordinals and the semantics of comparison classes. Ms.

Carrara, M., Arapinis, A., and Moltmann, F. (2016). *Unity & Plurality: Logic, Philosophy, and Linguistics*. Oxford University Press.

Chierchia, G. (1984). *Topics in the Syntax and Semantics of Infinitives and Gerunds*. PhD thesis, University of Massachusetts at Amherst.

Chierchia, G. (1998). Reference to kinds across languages. *Natural Language Semantics*, 6:339–405.

Chihara, C. S. (1990). *Constructibility and Mathematical Existence*. Oxford University Press.

Chihara, C. S. (2004). *A Structural Account of Mathematics*. Clarendon Press.

Dummett, M. (1973). *The Justification of Deduction*. Oxford University Press.

Dummett, M. (1991). *Frege: Philosophy of Mathematics*. Duckworth.

Felka, K. (2014). Number words and reference to numbers. *Philosophical Studies*, 168:261–268.

Field, H. (1980). *Science Without Numbers*. Princeton University Press.

Frana, I. (2006). The *de re* analysis of concealed questions: A unified approach to definite and indefinite concealed questions. In Gibson, M. and Howell, J., eds., *Proceedings of SALT 16*.

Frege, G. (1884). *Grundlagen der Arithmetik*.

Frege, G. (1903). *Grundgesetze der Arithmetik II*. Olms.

Frege, G. (1951). On concept and object. *Mind*, 60(238):168–180.

Geurts, B. (2006). Take 'five'. In Vogleer, S. and Tasmowski, L., eds., *Non-Definiteness and Plurality*, pages 311–329. Benjamins.

Greenberg, B. (1977). A semantic account of relative clauses with embedded question interpretations. Ms., University of California, Los Angeles.

Hale, B. (1987). *Abstract Objects*. Basil Blackwell.

Hale, B. (1994). Singular terms. In *The Philosophy of Michael Dummett*, pages 17–44. Springer.

Hale, B. (2016). Definitions of numbers and their applications. In Ebert, P. and Rossberg, M., eds., *Abstractionism*. Oxford University Press.

Hale, B. and Wright, C. (2001). *The Reason's Proper Study: Towards a Neo-Fregean Philosophy of Mathematics*. Oxford University Press.

Heck, R. (2011). *Frege's Theorem*. Clarendon Press.

Heim, I. (1979). Concealed questions. In Bauerle, R., Egli, U., and von Stechow, A., eds., *Semantics from Different Points of View*, pages 51–60. Springer.

Hellman, G. (1989). *Mathematics without numbers: Towards a modal-structural interpretation*. Clarendon Press.

Hellman, G. (1998). Maoist mathematics? *Philosophia Mathematica*, 6(3): 334–345.

Higgins, R. (1973). *The Pseudo-cleft Construction in English*. Garland.

Hodes, H. (1984). Logicism and the ontological commitments of arithmetic. *The Journal of Philosophy*, 81:123–149.

Hofweber, T. (2005). Number determiners, numbers, and arithmetic. *The Philosophical Review*, 114:179–225.

Hofweber, T. (2007). Innocent statements and their metaphysically loaded counterparts. *Philosopher's Imprint*, 7:1–33.

Hofweber, T. (2014). Extraction, displacement, and focus. *Linguistics and Philosophy*, 37(3):263–267.

Hofweber, T. (2016). *Ontology and the Ambitions of Metaphysics*. Oxford University Press.

Horn, L. (1972). *On the Semantic Properties of Logical Operators*. PhD thesis, University of California, Los Angeles.

Kennedy, C. (2012). Adjectives. In Russell, G. and Graff-Fara, D., eds., *Routledge Companion to the Philosophy of Language*. Routledge.

Kennedy, C. (2015). A "de-Fregean" semantics (and neo-Gricean pragmatics) for modified and unmodified numerals. *Semantics and Pragmatics*, 8(10):1–44.

Kennedy, C. and McNally, L. (2005). Scale structure, degree modification, and the semantics of gradable predicates. *Language*, 81:345–381.

Kennedy, C. and McNally, L. (2010). Color, context, and compositionality. *Synthese*, 174(1):79–98.

Krifka, M. (1989). Nominal reference, temporal constitution, and quantification in event semantics. In von Bentham, J., Bartsch, R., and von Emde Boas, P., eds., *Semantics and Contextual Expressions*. Foris.

Krifka, M. (1990). Boolean and non-boolean *and*. In Kalman, L. and Polos, L., eds., *Papers from the Second Symposium on Logic and Language*. Akademmiai Kiado.

Krifka, M., Pelletier, F., Carlson, G., ter Meulen, A., and Link, J. (1995). Genericity: An introduction. In Carlson, G. and Pelletier, F., eds., *The Generic Book, The University of Chicago Press, Chicago*, pages 1–124.

Landman, F. (2003). Predicate-argument mismatches and the adjectival thoery of indefinites. In Coene, M. and D'hulst, Y., eds., *NP to DP*. John Benjamins.

Landman, F. (2004). *Indefinites and the Type of Sets*. Blackwell.

Leng, M. (2005). Revolutionary fictionalism: A call to arms. *Philosophia Mathematica*, 13(3):277–293.

Lewis, D. (1986). *On the plurality of worlds*, volume 322. Oxford Blackwell.

Link, G. (1983). The logical analysis of plurals and mass terms: A lattice-theoretic approach. In Bäuerle, R., Schwarze, C., and von Stechow, A., eds., *Meaning, Use, and Interpretation of Langauge*, pages 303–323.

Link, G. (1998). *Algebraic Semantics in Language and Philosophy*. SCLI Publications.

Linnebo, Ø. (2009). The individuation of the natural numbers. In Bueno, O. and Linnebo, Ø., eds., *New Waves in Philosophy of Mathematics*, pages 220–238. Palgrave-MacMillan.

McNally, L. and de Swart, H. (2011). Inflection and derivation: How adjectives and nouns refer to abstract objects. In *Proceedings of the 18th Amsterdam Colloquium*, pages 425–434.

Mikkelsen, L. (2005). *Copular Clauses: Specification, Predication, and Equation*. Benjamins.

Mikkelsen, L. (2011). Copular clauses. In von Heusinger, K., Maienborn, C., and Portner, P., eds., *Semantics: An International Handbook of Natural Language Meaning*. de Gruyter.

Moltmann, F. (2008). Intensional verbs and their intentional objects. *Natural Language Semantics*, 16(3):239–270.

Moltmann, F. (2013a). *Abstract objects and the semantics of natural language*. Oxford University Press.

Moltmann, F. (2013b). Reference to numbers in natural language. *Philosophical Studies*, 162:499–536.

Moltmann, F. (2016). Plural reference and reference to a plurality. In M. Carrara, A. A. and Moltmann, F., eds., *Unity & Plurality: Logic, Philosophy, and Linguistics*. Oxford University Press.

Moltmann, F. (2017). Number words as number names. *Linguistics and Philosophy*, 40:331–345.

Nathan, L. (2006). *On the Interpretation of Concealed Questions*. PhD thesis, Massachussets Institute of Technology.

Nutting, E. S. (2018). Ontological realism and sentential form. *Synthese*, 195(11):5021–5036.

Oliver, A. and Smiley, T. (2013). *Plural Logic: Revised and Enlarged*. OUP Oxford.

Partee, B. (1986a). Ambiguous pseudoclefts with unambiguous *be*. In Bergman, S., Choe, J., and McDonough, J., eds., *Proceedings of the Northwestern Linguistics Society 16*. GLSA.

Partee, B. (1986b). Noun phrase interpretation and type-shifting principles. In Groenendijk, J., de Jongh, D., and Stokhof, M., eds., *Studies in Discourse Representation Theory and the Theory of Generalized Quantifiers*. Foris.

Partee, B. (2004). *Compositionality in Formal Semantics*. Blackwell Publishing.

Partee, B. and Borschev, V. (2012). Sortal, relational, and functional interpretations of nouns and russian container constructions. *Journal of Semantics*, 29:445–486.

Partee, B. and Rooth, M. (1983). Generalized conjunction and type ambiguity. In Bauerle, R., Schwarze, C., and von Stechow, A., eds., *Meaning, Use, and Interpretation of Language*, pages 361–383. De Gruyter.

Partee, B. H. (1992). semantic type. *Computational linguistics and formal semantics*, page 97.

Rett, J. (2008). *Degree Modification in Natural Language*. PhD thesis, Rutgers University.

Romero, M. (2005). Concealed questions and specificational subjects. *Linguistics and Philosophy*, 28(6):687–737.

Rothstein, S. (2010). Counting, measuring and the semantics of classifiers. *Baltic International Handbook of Cognition, Logic and Communication*, 6.

Rothstein, S. (2013). A Fregean semantics for number words. In *Proceedings of the 19th Amsterdam Colloquium*, pages 179–186. Universiteit van Amsterdam, Amsterdam.

Rothstein, S. (2016). Counting and measuring: a theoretical and crosslinguistic account. *Baltic International Yearbook of Cognition, Logic and Communication*, 11(1):8.

Rothstein, S. (2017). *Semantics for Counting and Measuring*. Cambridge University Press.

Schiffer, S. R. (2003). *The Things We Mean*. Oxford University Press.

Schlenker, P. (2003). Clausal equations. *Natural Language and Linguistic Theory*, 21:157–214.

Schwartzkopff, R. (2016). Singular terms revisited. *Synthese*, 193(3):909–936.

Schwarzschild, R. (2005). Measure phrases as modifiers of adjectives. *Recherches Linguistiques de Vincennes*, 34:207–228.

Scontras, G. (2014). *The Semantics of Measurement*. PhD thesis, Harvard University.

Shapiro, S. (1997). *Philosophy of Mathematics: Structure and Ontology*. Oxford University Press.

Shapiro, S., Snyder, E., and Samuels, R. (ms.). What's wrong with Hofweber's nominalism.

Sharvy, R. (1980). A more general theory of definite descriptions. *The Philosophical Review*, 89:607–624.

Snyder, E. (2017). Numbers and cardinalities: What's really wrong with the easy argument? *Linguistics and Philosophy*, 40:373–400.

Snyder, E. (2020). Counting, measuring, and the fractional cardinalities puzzle. *Linguistics and Philosophy*.

Snyder, E. and Barlew, J. (2016). The universal measurer. In *Proceedings of Sinn und Bedeutung 20*.

Snyder, E. and Barlew, J. (2019). How to count $2\frac{1}{2}$ oranges. *Australasian Journal of Philosophy*.

Snyder, E., Samuels, R., and Shapiro, S. (2018a). Neologicism, Frege's Constraint, and the Frege-Heck condition. *Noûs*.

Snyder, E., Samuels, R., and Shapiro, S. (ms.). Hofweber's nominalist naturalism.

Snyder, E., Samuels, R., and Shaprio, S. (2019). Hale's argument from transitive counting. *Synthese*, pages 1–29.

Snyder, E. and Shapiro, S. (2016). Frege on the reals. In Ebert, P. and Rossberg, M., eds., *Essays on Frege's Basic Laws of Arithmetic*. Oxford University Press.

Snyder, E. and Shapiro, S. (2020). Mereological singularims and paradox. *Erkenntnis*.

Snyder, E., Shapiro, S., and Samuels, R. (2018b). Cardinals, ordinals, and the prospects for a Fregean foundations. In *Royal Institute of Philosophy Supplements: RIP Metaphysics*. Cambridge University Press.

Wright, C. (1983). *Frege's Conception of Numbers as Objects*. Aberdeen University Press.

Wright, C. (2000). Neo-Fregean foundations for real analysis: Some reflections on Frege's constraint. *Notre Dame Journal of Formal Logic*, 41:317–334.

Wright, C. et al. (1999). Is Hume's principle analytic? *Notre Dame Journal of Formal Logic*, 40(1):6–30.

The Philosophy of Mathematics

Penelope Rush

University of Tasmania

From the time Penny Rush completed her thesis in the philosophy of mathematics (2005), she has worked continuously on themes around the realism/anti-realism divide and the nature of mathematics. Her edited collection *The Metaphysics of Logic* (Cambridge University Press, 2014), and forthcoming essay 'Metaphysical Optimism' *(Philosophy Supplement)*, highlight a particular interest in the idea of reality itself and curiosity and respect as important philosophical methodologies.

Stewart Shapiro

The Ohio State University

Stewart Shapiro is the O'Donnell Professor of Philosophy at The Ohio State University, a Distinguished Visiting Professor at the University of Connecticut, and a Professorial Fellow at the University of Oslo. His major works include *Foundations without Foundationalism* (1991), *Philosophy of Mathematics: Structure and Ontology* (1997), *Vagueness in Context* (2006), and *Varieties of Logic* (2014). He has taught courses in logic, philosophy of mathematics, metaphysics, epistemology, philosophy of religion, Jewish philosophy, social and political philosophy, and medical ethics.

About the Series

This Cambridge Elements series provides an extensive overview of the philosophy of mathematics in its many and varied forms. Distinguished authors will provide an up-to-date summary of the results of current research in their fields and give their own take on what they believe are the most significant debates influencing research, drawing original conclusions.

Cambridge Elements ☰

The Philosophy of Mathematics